how to
know the

freshwater
fishes

The **Pictured Key Nature Series** has been published since 1944 by the Wm. C. Brown Company. The series was initiated in 1937 by the late Dr. H. E. Jaques, Professor Emeritus of Biology at Iowa Wesleyan University. Dr. Jaques' dedication to the interest of nature lovers in every walk of life has resulted in the prominent place this series fills for all who wonder **"How to Know."**

John F. Bamrick and Edward T. Cawley
Consulting Editors

The Pictured Key Nature Series

How to Know the

AQUATIC PLANTS, Prescott	**ROCKS AND MINERALS,** Helfer
BEETLES, Jaques	**SEAWEEDS,** Abbott-Dawson, Second Edition
BUTTERFLIES, Ehrlich	**SPIDERS,** Kaston, Third Edition
CACTI, Dawson	**SPRING FLOWERS,** Cuthbert, Second Edition
EASTERN LAND SNAILS, Burch	**TAPEWORMS,** Schmidt
ECONOMIC PLANTS, Jaques, Second Edition	**TREMATODES,** Schell
FALL FLOWERS, Cuthbert	**TREES,** Miller-Jaques, Third Edition
FERNS, Mickel	**TRUE BUGS,** Slater-Baranowski
FRESHWATER ALGAE, Prescott, Second Edition	**WATER BIRDS,** Jaques-Ollivier
FRESHWATER FISHES, Eddy-Underhill, Third Edition	**WEEDS,** Wilkinson-Jaques, Second Edition
GRASSES, Pohl, Third Edition	**WESTERN, TREES,** Baerg, Second Edition
GRASSHOPPERS, Helfer, Second Edition	
IMMATURE INSECTS, Chu	
INSECTS, Bland-Jaques, Third Edition	
LAND BIRDS, Jaques	
LICHENS, Hale	
LIVING THINGS, Jaques, Second Edition	
MAMMALS, Booth, Third Edition	
MARINE ISOPOD CRUSTACEANS, Schultz	
MITES AND TICKS, McDaniel	
MOSSES AND LIVERWORTS, Conard, Second Edition	
NON-GILLED FLESHY FUNGI, Smith-Smith	
PLANT FAMILIES, Jaques	
POLLEN AND SPORES, Kapp	
PROTOZOA, Jahn, Jahn, Bovee, Third Edition	

how to
know the
freshwater
fishes

Third Edition

Samuel Eddy
formerly University of Minnesota

James C. Underhill
*James Ford Bell Museum of National History
and University of Minnesota*

The Pictured Key Nature Series
Wm. C. Brown Company Publishers
Dubuque, Iowa

Copyright © 1957 by H.E. Jaques

Copyright © 1969, 1978 by Wm. C. Brown Company Publishers

Library of Congress Catalog Card Number: 77—82895

ISBN 0—697—04750—4 (Paper)
ISBN 0—697—04751—2 (Cloth)

Printed in the United States of America
20   19   18   17   16

Contents

Preface

These keys for the identification of freshwater fishes of the United States were originally constructed by the late Professor Samuel Eddy for use in his ichthyology classes at the University of Minnesota. I originally had the opportunity to "test" the first keys, along with a number of fellow graduate students, in a year of weekly seminars dealing with fish. Later I had the pleasurable experience of seeing and participating in the preparation of the second edition. Just prior to his death Professor Eddy was contemplating the present third edition and he gave me his annotated copy. The key is largely his and its past success is a reflection of his dedication to the fishes and to his tremendous artistic talents.

The present edition attempts to bring the keys, nomenclature and distribution information for the freshwater fishes of the United States up to date as nearly as possible. In the past eight years many new species of freshwater fishes have been described. Also a number of new books dealing with the fishes of various states and the Provinces of Canada have been published providing us with a more detailed knowledge of the distribution of our freshwater fishes. A number of very important detailed research papers dealing with various genera have been published in which the relationships between the various species have been more clearly delineated. Much new information has come from the work of ichthyologists in the southern and western United States. A few additional exotic species have become established in the southern United States and a number of native species have been introduced intentionally or accidentally outside their former range.

Nancy Bagley and Louisa Schmid prepared the new illustrations. I am much indebted to them for their talents, for their suggestions, and their continued good humor during the preparation of the revision. Anne Underhill provided assistance with the manuscript and encouragement throughout these months. Jim Erickson read many portions of the manuscript and I have benefited from his criticisms, comments, and suggestions.

I wish to acknowledge the encouragement and many subtle contributions of my friends and associates; Jim Erickson, David Merrell, Magnus Olson, and William Schmid. My association with the James Ford Bell Museum of Natural History has been of great value and I wish to acknowledge the kind cooperation of Charles Huver, Curator of Fishes, and Harrison Tordoff, the museum director.

I am indebted to the many persons who offered helpful criticisms of the previous edition of the book and suggestions for improvements, especially George C. Becker, Thomas Buchanan, Glenn Clemmer, Joseph Eastman, Carter Gilbert, Edward Kott, Robert

Kuehne, Peter Moyle, and Thomas Todd. Reeve Bailey and Carl Hubbs have been an important influence on this and previous editions and their contributions have been of great significance and have been much appreciated by Professor Eddy and by me. I am very appreciative of the kindness of William R. Taylor in allowing me to modify his key to the genus *Noturus* and for permission to adapt his figures of the pectoral spines of several of the madtoms. Edward Kott provided me with up-to-date information on the lampreys for which I am most grateful. The key to the coregonine fishes is adapted in part from the key prepared by John Parsons and Thomas Todd, and I am much indebted to them for their generosity. Glenn Clemmer kindly provided information on several species of the genus *Hybopsis.*

Brooks Burr provided data on the genus *Campostoma.*

As noted in previous editions of this book there are many biologists who have contributed fishes to the James Ford Bell Museum collections and to all these persons I express my sincere thanks. Recently the following biologists have provided specimens on permanent loan and for use in the present edition; Eugene Beckman, George Becker, Carl Bond, Thomas Buchanan, Neil Douglas, David Etnier, Carter Gilbert, Edward Kott, Robert Kuehne, Peter Moyle, John Peterka, William Palmer, Royal Suttkus, and Thomas Todd. Their contributions have been of immense value to me and to the museum.

James C. Underhill

Some Facts About Fishes

Fishes are the largest group of vertebrates and are represented by four separate major subdivisions or classes: the Agnatha or Cyclostomata, the Placodermi, the Chondrichthyes, and the Osteichthyes. Of these groups all but the Placodermi are represented by living forms. The first vertebrates known from the fossil record were the Ostracoderms, a diverse group of jawless fishes with bony plates covering the body. Ostracoderm fossils first appear in deposits of Ordovician age, approximately 450 million years ago. The lampreys and the marine hagfishes are the modern-day survivors of this once numerous group. The first jawed vertebrates were the Placodermi or plate-skinned fishes, which in turn gave rise to the Osteichthyes or bony fishes and the Chondrichthyes or cartilaginous fishes (sharks, skates, and rays). All four groups were present in the Devonian period (405 to 350 million years ago), which is referred to as the "Age of Fishes." Our modern fishes are the survivors of hundreds of millions of years of evolution from these ancient forms. A few such as the lampreys have changed little in the past 300 million years. Others, especially members of the Osteichthyes, have undergone an explosive radiation into the marine and freshwater habitats. Of the approximately 21,000 species of living fishes 97 percent are bony fishes.

The modern bony fishes are represented by two strikingly different groups (sub-classes), the Sarcopterygii (flesh or lobed-finned fishes) and the Actinopterygii (ray-finned fishes). The latter comprise over 99 percent of the living bony fishes. Three sub-groups (superorders) make up the ray-finned fishes; the primitive Chondrostei (paddlefishes and sturgeons) and Holostei (garfishes and bowfin) with approximately 33 species and the Teleostei with some 20,000 species. The Teleostei is the most diverse group of vertebrates living today and has been the dominant group of fishes for almost 100 million years. Their numbers equal the total numbers of living mammals, birds, reptiles, and amphibians combined. The teleosts comprise at least 27 different orders and 390 families, the majority of which are marine. For example, in this book only 39 families of teleosts are listed as being present in fresh water. Of these, eight represent families with one or more marine species that occasionally enter fresh water. Therefore, while the freshwater fish fauna is quite diverse, it gives only a hint of the tremendous evolutionary and ecological diversity of the teleosts.

Most of the fishes in our inland waters are restricted by their physiology to life in fresh water and cannot survive in the sea. Of our native fishes twelve families have such physiological restrictions, including such families as the minnows (Cyprinidae), suckers

(Catostomidae), catfishes (Ictaluridae), sunfishes (Centrarchidae), and perch and darters (Percidae). These 6 families comprise over seventy-five percent of our freshwater fish fauna.

Some freshwater families have species which can tolerate low salinities and these species often invade brackish waters at the mouths of rivers, but never venture far into the sea. Some of the saltwater fishes frequently invade the fresh water at the mouths of rivers and some may penetrate upstream for several hundred miles. In this book some of the marine fishes which commonly invade fresh water are included, but the number occasionally entering fresh water is so great that it would make the keys too cumbersome to include all marine fishes that have been reported from fresh water. Fishes which spend most of their lives in fresh water but go to sea to spawn, are known as *catadromous* fishes. The only common catadromous fish in the United States is the American eel. Fishes which spawn in fresh water, but spend most of their lives in the sea, are known as *anadromous* fishes. Many fishes, such as, the Pacific salmon, some shad and smelt spend most of their lives in the sea, but regularly enter fresh water to spawn. These species are considered as freshwater species in this book. Some freshwater fishes, such as the trout, have anadromous races which commonly go to sea, but return to fresh water to spawn.

The freshwater fishes of the United States do not reach the enormous size attained by many of the marine fishes or by some of the freshwater fishes in other parts of the world. A weight of several hundred pounds may be reached by some catfishes and the lake sturgeon. The anadromous white sturgeon of the Pacific northwest has been reported weighing over 1,000 pounds. Several of our freshwater fishes are quite minute seldom exceeding an inch in length such as gambusia or the mosquitofish, the pygmy sunfish, and the least killifish.

DISTRIBUTION AND CONDITIONS FOR EXISTENCE

The distribution of fishes is usually determined by stream systems as land divides often constitute an effective barrier. The greatest separation of American fishes is caused by the continental divide which rather effectively has separated the fishes of the Pacific drainage from those of the Atlantic drainage. In most cases entirely different species and even genera occur on the two sides of the divide. A few species have crossed apparently at the narrow divide between the headwaters of the Missouri and Columbia Rivers. The Arctic and the Great Lakes drainages have had many connections with the Mississippi drainage and, consequently, show many species common to both. Several fishes found in the Arctic drainage have penetrated into the northern part of the Mississippi drainage, but are probably restricted from going farther south because of suitable living conditions. The Atlantic drainage shows that many fishes from the Mississippi drainage have been able to cross the divide, but there are many species which are restricted to the streams of the Atlantic seaboard. In some cases these are continuous in the coastal streams along the Gulf of Mexico.

There are many cases of isolated stream systems containing endemic species, such as, stream systems of the southwestern desert which have lost all connections with other basins and empty into lakes without any outlets. Also coastal rivers, such as the Sacramento and many rivers in southeastern United States, are isolated and have developed partially endemic fish faunas.

Many fishes show individual preferences for certain water conditions and are to a certain extent restricted in their distribution by these conditions. Some fishes, such as members of the salmon family, are restricted to cold waters and will not be encountered in regions where there are no waters within their optimum range of temperature. Other fishes prefer warmer waters, such as black basses and sunfishes, and thrive best in waters which reach temperatures above 75°. The various ciscoes, the Great Lakes whitefish, the lake trout, and a few other fishes are restricted to lake waters and avoid streams. Other fishes prefer running waters and are more likely to be found in rivers. Certain darters are found only in small swift streams. Fishes, such as the larger catfishes, are more likely to be found in the larger and more placid rivers. Land barriers between stream systems are not the only condition limiting the range of a fish, as the proper habitat conditions for that particular species must also be present.

Fishes depend on many other conditions for their existence, but fortunately many of these conditions, such as food, are ample in most fresh waters. If the salt content is too great, as in Great Salt Lake, it will prevent any fish from living there, but in most waters the salt content is within the tolerance for most fishes. The carbonate or lime concentration, although important to the growth of food organisms, is usually within the tolerance of fishes. Proper spawning beds are a very important factor and often form a limiting factor as most fishes need certain depths, bottom types and water temperatures for spawning, and without these conditions they cannot maintain the species.

A necessary factor in the existence of fishes is the presence of sufficient oxygen for respiration. Most fishes obtain their oxygen from that dissolved in the water which passes over their gill filaments and such fishes cannot live when it falls below a certain concentration. However, a few *physostomic* fishes can use atmospheric oxygen by means of their air bladder when that in the water becomes insufficient for gill respiration.

The *air bladder,* a membranous sac filled with gases and just above the alimentary tract usually functions as a hydrostatic organ regulating the buoyancy of a fish. In the few physostomic fishes it retains an opening into the pharynx although closed in most *physoclistic* fishes. The physoclistic fishes may use some of the oxygen in their air bladders in emergency. Some also use their air bladders for sound production and perhaps for sound perception.

Some shallow waters in the north may develop an insufficient amount of oxygen during the winter. Much of the oxygen in standing water originates from the oxygen produced by plants during photosynthesis. When sunlight is cut off by the snow on the ice, photosynthesis stops, and the shallow lakes may lose most or all of their dissolved oxygen. Streams obtain their oxygen largely from the atmosphere by diffusion and as a result of the constant mixing that occurs in flowing water. Consequently, streams usually do not show as much reduction in oxygen during the winter. An important contributing factor to oxygen reduction is the decomposition of organic matter. An abundance of organic matter, such as a heavy weed crop or domestic sewage, may through the oxidative processes of decomposition result in a depletion of the dissolved oxygen. Such depletions may occur in both lakes and streams.

In many of the deep northern lakes, the cold water stratifies in summer and remains below the warmer surface water and is too deep for photosynthetic activity. If considerable organic matter has settled into this deep water, oxidation may soon deplete the oxygen from the lower levels and cause the fishes to be confined to the upper levels. Only those deep lakes which are not fertile enough to produce much plant life have sufficient oxygen to maintain fishes in their lower levels. Large lakes with strong currents may keep the water sufficiently stirred to prevent any stagnation.

ACTIVITIES

Most fishes feed on or close to the bottom and hence are restricted to water where they can always reach the bottom. A few fishes are pelagic and live in the deep open waters of large lakes. These feed chiefly on plankton or other fishes which are in turn plankton feeders. Fishes exhibit all sorts of feeding habits. Young fishes when first hatched usually start feeding on the minute crustacea which swarm in the shallow water. Many soon turn to small insects and fry of other fishes. Many of the minnows, bullheads, and other rough fishes consume large quantities of plant food. A few fishes are plankton feeders, possessing fine gill rakers by which they strain out the tiny crustacea and other planktonic forms which swarm through the open waters of all lakes. The game fishes are most predaceous, feeding on smaller fishes and on all sorts of other aquatic animal life. Thus long chains of food habits are established. The forage fishes feeding on plants and on plankton, furnish food for the predaceous fishes which top the chain. Suckers sweep over the bottom with their sucker mouths utilizing anything that is edible. In between are the insectivorous fishes, such as crappies, sunfishes, and perch, feeding mainly on the smaller animal life, but occasionally feeding on small fishes and in turn sometimes eaten by the larger game fishes.

Fishes exhibit definite activity periods as do most animals. Some are diurnal and become active after sunup. Others are nocturnal and are most active at night. Night feeders usually have a keen sense of smell and taste by which they partly or wholly locate their food. Diurnal fishes depend more on sight than taste and smell to locate their food and some such as the pike, use sight almost entirely.

Many fishes are gregarious and tend to keep together in "schools." Others, such as the adult pike and black basses, are solitary. The black basses for the first six months of their lives are gregarious, but they soon separate and each male more or less selects his own territory which he defends against all invaders. The related sunfishes remain gregarious, and even when spawning are so sociable that they may put their nests as close together as possible. Bullheads are gregarious and swarm in schools.

Fishes exhibit several types of daily and annual movements. The spawning runs of many are well-known. The suckers and the walleyes follow definite paths to their spawning beds at the start of each spring. The mad crowding rush of the Pacific salmon, smelt, shad, and many other anadromous fishes to their spawning grounds are well-known classic examples. Less spectacular are the spawning runs of many of our freshwater fishes, such as those of the suckers.

We are just beginning to learn about the daily movements of many of our common fishes. Pike move into the shallow waters during the day to feed and at evening pass to the deeper waters outside of the weed beds to spend the night. On the other hand, the walleyes move inshore at sundown and spend the night in shallow water moving outside the weed beds during the early morning to spend the day. Perch and sunfishes also exhibit similar daily movements. Each kind of fish seems to have its own pattern of activity and the pattern may vary with the age of the fish.

REPRODUCTION

Fishes are usually prolific breeders, producing enormous numbers of eggs which compensates for the high mortality prior to hatching and in the fry and fingerlings. The number of eggs produced per fish may vary from 15 to 20 in the live-bearers to as high as a million in the carp and eel. Most fishes produce eggs which are fertilized externally and hatch after they have been laid. A few species such as members of the family Poeciliidae, are live-bearers. In these fish the eggs are retained in the ovary and fertilized. Ovulation is delayed until the embryos are ready to hatch and the young fish pass into the oviduct and out of the female. The males of live-bearers have a modified anal fin, the gonopodium, used to inseminate the female. In these fishes the number of eggs produced is few, the probability of survival of each is quite high.

The majority of our freshwater fishes are egg-laying and have developed two methods for developing the eggs. One method, and perhaps the most common, is that of depositing the eggs at random on suitable but unprepared spawning beds. The eggs are fertilized as they are laid by one or more attending males and are left to develop and hatch without any further care. These random spawners produce enormous numbers of eggs, often many thousands or more. A number of our fishes, such as sunfishes and catfishes, are nest builders and the males generally prepare a nest, usually a cleared depression where the female deposits the eggs which are then guarded by the male who also guards the young fry for some time afterwards. These fishes usually produce only a few thousand eggs, and the chance for survival is much greater than in the random spawners. There are a number of fishes that have partial nesting habits, depositing the eggs in a prepared nest and even guarding the eggs but giving no care to the young. Many minnows prepare nests and some even guard the eggs. Many trout make some preparation covering their eggs with gravel and then leaving the eggs to shift for themselves.

Members of the family Cichlidae, including many introduced species, exhibit very elaborate courtship patterns. Some members of the family are mouthbreeders, all are nest builders. The male and/or female guard the nest and eggs from potential predators, clean the eggs and move them occasionally. The mouthbreeders brood the fertilized eggs in their mouth until the eggs hatch. The female picks up the eggs as soon as they are fertilized and may carry them in her buccal cavity for over two weeks. She continues to guard her brood and if they are disturbed the offspring swim back into their mother's mouth. In some species the male carries the young. Some species of mouthbreeders may carry their young to a sheltered "brooding ground" where the fry are released. These nurseries may serve to reduce the predation on the young in more open areas of the lake.

Structure of a Fish*

In order to identify a fish it is necessary to know something about the structure of a fish especially those parts used in classification. The shape of fishes varies greatly. Many have slender streamlined bodies, but others develop thick heavy bodies, fitting almost every conceivable dimension. Some may be very long and cylindrical as in the eel, others are compressed laterally and are deep vertically as in the sunfishes. Proportions vary greatly. Some fishes have large wide heads and small slender bodies, while others may have small heads with wide heavy bodies.

The general terms of anatomical dimensions apply to fishes the same as to other animals. *Anterior* refers to before or to the front end or part of the body or structure. *Posterior* refers to behind or to the hind end or part of the body or structure. *Dorsal* refers to the back or upper surface. *Ventral* refers to the under part or lower surface. *Lateral* means the sides or toward the sides. *Medial* refers to the central part or middle of the body or structure.

The body of a fish is divided into three regions, consisting of *head, trunk,* and *caudal* regions (Fig. 1). No neck is present, although the region of the back just behind the head is called the *nape*. The head is that part extending to the posterior edge of the gill cover or *opercle*. The trunk is the region from the edge of the *opercle* to the anus. Several areas may be found in the trunk. The *pectoral* (shoulder) area is that just behind the opercle and includes the *humeral* area which is the area just above the base of the pectoral fin. The *abdomen* or *belly* is the extreme ventral portion between the pectoral fins and the anus. The *thorax* or *breast* is the ventral area immediately in front of the pectoral fins.

The *tail* or *caudal* region (Fig. 1) is the region from behind the anus extending to the caudal fin, and is not the caudal or tail fin. The more or less slender part of the caudal region behind the anal or dorsal fin (whichever extends farthest back) and extending to the base of the caudal fin is the *caudal peduncle*. The *anus* (Fig. 2) is the posterior opening of the digestive tract and is adjacent to the openings of the urogenital tracts. The general area of the anus is often swollen.

Fishes possess several kinds of fins, which are usually membranous structures supported by rays or spines. Rays are modified into soft and hard rays. *Soft rays* (Fig. 1) are slender flexible structures composed of many bony joints and are typically split or divided at their outer ends. The soft rays at the front of a fin

*The cyclostomes represented by the lampreys (page 26) form a type whose structure is more primitive in many ways than that of the other fishes and, consequently, the following description applies to the bony fishes.

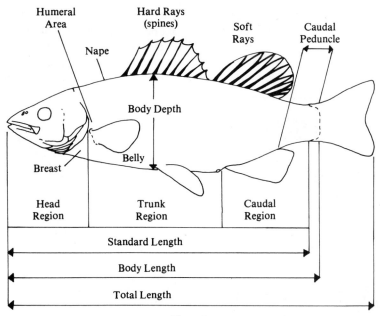

Figure 1.

are usually short and are not divided at their tips and are known as *rudimentary soft rays* (Fig. 2). When counts of the fin rays are made, the short rudimentary rays are not included, but the long unbranched ray usually found at the front of the dorsal and anal fins is usually included in the count. The last soft ray of both dorsal and anal fins is often split almost to the base and may be mistaken for two rays.

In a few fishes, such as the catfishes and the carp, groups of soft rays may fuse into a stiff spine-like structure known as *hard rays*. These are usually barbed. If their membranous covering is removed their jointed structure will be detected. *True spines* (Fig. 1) are stiff rays ending in sharp points and do not show a jointed structure.

The median or unpaired fins of a fish consist of the dorsal, caudal, and anal fins. The *dorsal fin* (Fig. 2) extends along the middle of

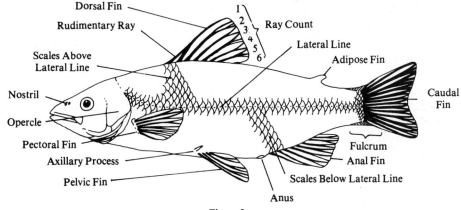

Figure 2.

the back and may be divided into several parts, the anterior portion often being spiny. The tail terminates in the *caudal fin* which has developed several types. Primitive fishes or relics of ancient groups have a *heterocercal* type (Fig. 3) in which the vertebral column ex-

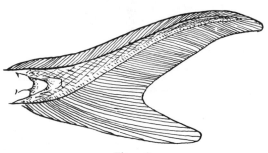

Figure 3.

tends out into the upper lobe of the fin. A modification of this type (See Fig. 14.) occurs in the families Amiidae and Lepisosteidae, where the young are hatched with typical heterocercal fins, but lose the upper lobe as they grow. Most fishes have a *homocercal* type (Fig. 4) of caudal fin where the vertebral col-

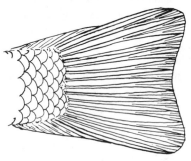

Figure 4.

umn ends at the base of the fin. This type may be forked, rounded, or square. The caudal fin is composed of soft rays with rudimentary rays on each side. The term *fulcrum* (Fig. 2) applies to the swollen area above and below the base of the caudal fin produced by the continuation of the rudimentary rays.

The *anal fin* (Fig. 2) is a median ventral fin located just posterior to the anus. It may be

composed of both spines and soft rays. The shape of the anal and dorsal fins is usually not highly variable, but sometimes one or both of these fins may assume a *falcate* (sickle-shape) form (Fig. 5) with an "S"-shaped edge.

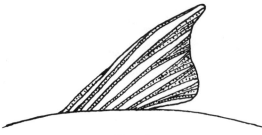

Figure 5.

Fins corresponding to arms and legs are present on most fishes, although one or both pairs are lost in a few fishes. The anterior pair of fins are the *pectoral fins* (Fig. 2) located laterally on the shoulder girdle just back of the *opercle*. The *pelvic fins* (Fig. 2) are typically located just anterior to the anus, but in many fishes they move forward. When the pelvic fins are near the anus, they are termed abdominal in position (See Fig. 22.), but when they are near or under the pectorals they are termed thoracic in position. (See Fig. 23.) In some species they may be anterior to the pectoral fins and are termed jugular in position. In many fishes slender ridges or structures known as *axillary processes* (Fig. 2) are found in the angles at the base of the pelvic fins.

Another type of fin found in some fishes is the *adipose fin* (Fig. 2) characteristic of trout, catfishes and several other groups. This is a small median fin behind the dorsal fin distinguished by being a soft fleshy structure without any rays or spines.

The body of a fish is ordinarily more or less covered with *scales:* sometimes the scales are so small they can barely be detected. Areas without scales are usually said to be naked. A few fishes have lost their scales entirely. Scales are of several types. Several of the most primitive bony fishes possess hard rhomboid or diamond-shaped scales which do not overlap and are called *ganoid scales* (Fig. 6). Many primitive fishes also have heavy bony plates on the body covering the heads. In the teleost fishes the bony plates of the head have been incorporated with the internal skeleton and are not easily discernable.

Figure 7.

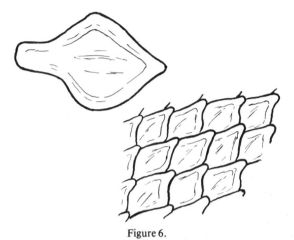

Figure 6.

Two types which are really modifications of the same scale are found in the higher bony fishes. Each scale is a thin shingling disc of bone with the exposed part covered by a very thin skin. The scale is formed by concentric layers of bone (circuli) which are laid down at the margin of the scale as the fish grows. During the winter when growth ceases or is retarded, the scale may suffer some reabsorption at the margin. When growth is resumed in the spring this causes a distinct mark known as an annulus which is used to determine the age of the fish. Ridges appearing as lines radiate out from the center of the scale and are known as *radii*. The simple smooth scales are the *cycloid type* (Fig. 7). *Ctenoid scales* (Fig. 8) are similar, but are differentiated by tiny spines covering the exposed portion. Frequently the scale must be removed and magnified to determine the structure. Scales are usually restricted or may be absent on the head. In some fishes the scales may be absent from the nape, belly, and breast. The scales are counted on various

parts of the body for identification (see page 12 for various counts). Some scales are modified as the enlarged scales on the mid-belly of some darters. Many darters and some killifishes have an enlarged *humeral scale* located just behind the opercle and above the base of the pectoral fin. In some this is not a true scale but part of the exposed shoulder girdle. The thin skin of fishes contains numerous mucous glands which keep the skin covered with a slime which is protective, preventing bacteria and moulds from infecting the delicate skin.

Figure 8.

The head of a fish includes the gill region which corresponds to the neck and throat region of higher animals. The fleshy part of the head before the eye and above the mouth is the *snout* (Fig. 9). Its length is determined as the distance from the front or tip to the anterior margin of the orbit. This part contains the *nostrils* which are primarily a pair of blind pits and function only as smell organs. Each nostril aperture is divided by a flap or fleshy partition into an incurrent and excurrent opening

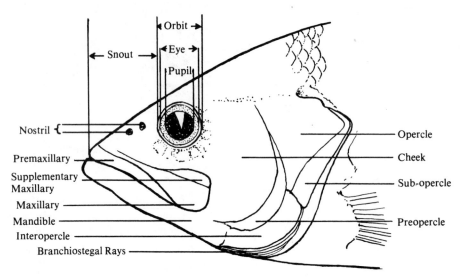

Figure 9.

(Fig. 9). The upper jaws under the snout are formed of bones covered by skin and a thin layer of flesh except in a few fishes which develop fleshy lips. The upper jaw (Fig. 9) consists of several pairs of bones. The front and outer pair is the *premaxillaries* which may be separated from the snout by a distinct groove (See Fig. 181.) in which case they are termed *protractile*. If a bridge of flesh crosses the groove and connects the premaxillaries to the snout (See Fig. 182.), they are termed *non-protractile*. The *maxillary* (Fig. 9) is on each side of the upper jaw and above and behind, but often parallel to the premaxillary. A splint-like supplementary maxillary may be applied to the upper edge of the maxillary. The posterior end of the maxillary usually marks the end of the jaw, and its position in relation to the eye or orbit is often used in identification.

The lower jaw consists of several bones, the most important consisting of the *dentaries* which usually bear teeth. In a few primitive fishes, a prominent shield-like bone, the *gular plate* (Fig. 10), lies between the right and left jaws. The length of the lower jaw varies in different species; in some it may protrude beyond the upper jaw while in others it may be equal or

may be shorter or *inferior*. The forward angle of the mandible forms the *chin*.

Almost any bone in the mouth of fishes is capable of bearing teeth. The roof is formed by an unpaired median *vomer* on each side of which are *palatines* extending to the pterygoids. In the floor of the mouth a bump formed by the protrusion of the *hyoid* bone and frequently bearing teeth, forms the *tongue*. The mouth when approximately at the anterior end of the head is said to be *terminal*. If the snout extends

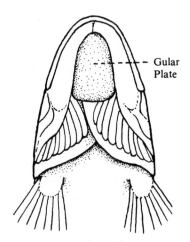

Figure 10.

considerably before the mouth, the mouth is said to be *sub-terminal*.

The *barbels* are thread-like structures on the head especially around the mouth of many fishes. These are prominent on such fishes as catfishes, but may be small bumps at the end of the maxillary of some minnows. (See Fig. 202.)

The eye of the fish lies within the *orbit* (Fig. 9). The external diameter of the orbit or the distance from rim to rim is often used as a comparative measurement. Behind the eye, the *cheek* (Fig. 9) is the fleshy area extending to the edge of the *preopercle* which is marked by a groove. The bony *opercle* or gill cover lies back of the cheek and marks the posterior border of the head. The opercle consists of the thinly covered *opercular* bone below which are the subopercular and interopercular bones. The space under the eye and extending to the maxillary bone is the *sub-orbital* region.

The *gill* or *branchiostegal membrane* (Fig. 9) is a thin membrane connecting the lower part of the opercle with the throat or with the opposite membrane. The membrane may form a close attachment with the throat or with the opposite membrane (Fig. 11), or it may extend far forward with a wide attachment leaving the anterior extension of the throat exposed as an *isthmus* (Fig. 12). The gill membrane is supported by a series of small slender bones

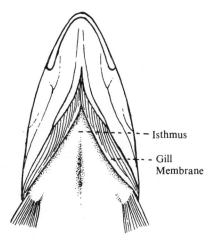

Figure 12.

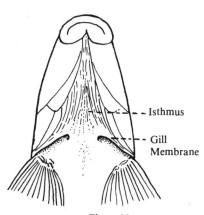

Figure 11.

known as the *branchiostegals* or *branchiostegal rays* (Fig. 9).

Under each opercle is the gill or *branchial chamber,* containing usually 4 sets of gills. Each set of gills consist of a pair of bony, flesh-covered *pharyngeal arches* supporting a double row of red gill filaments. These are the actual respiratory structures of fishes. On the inner surface of the gill arch is a row of finger-like structures (may be filamentous in some) which are the *gill rakers* and may serve to prevent any objects from entering the gill chambers from the throat. Fishes obtain oxygen from the water which enters through the mouth and passes out over the gills.

In some fishes a patch of rudimentary gill fragments known as *pseudobranchia* may be present on the inner surface of the opercular flap. The fifth gill or pharyngeal arches become modified and no longer bear gills in many fishes, but may develop tooth-like structures known as pharyngeal teeth. (See Figs. 33, 35.) These are well developed in suckers, minnows, and other fishes which may not have teeth in their mouths. The arrangement and number of the pharyngeal teeth in the minnows is a most important character used in identification.

Fishes possess an external set of sensory structures known as the *lateral line* system. The most conspicuous part of this system appears as the *lateral line* (Fig. 2) commonly seen on the sides of the trunk and tail regions. A pattern of pores related to this system can sometimes be traced over the head. The lateral line consists of an external row of pores, one on each scale, which open into a canal imbedded under the skin. The sensory endings of a branch of the 10th cranial nerve lie in this canal. Many functions have been assigned to this system, but the most recent findings indicate that it functions in receiving vibrations from objects thus enabling the fish to swim blindly without hitting objects and also helping in capturing prey.

The size of various structures are important characters used in fish identification. Individual fishes vary so much in size that actual measurements are of little value, consequently, comparative ratios are generally used. Hence, the number of times the eye goes into the length of the snout or the number of times the body depth or the length of the head goes into the standard length is more significant than the actual measurement. *Depth of body* (Fig. 1) is the greatest depth of body measured in a straight line from dorsal to ventral surface at right angles to the length.

The length of a fish is often considered as a straight line measurement from the tip of the jaws or the tip of the snout, if the snout extends beyond the mouth to the various parts. Dividers and a millimeter rule or dial calipers should be used for all small fishes. *Total length* (Fig. 1) is the distance to the extreme tip of the caudal fin. *Fork length* is the distance to the fork of the caudal fin. *Body length* (Fig. 1) is to the base of the caudal fin. *Standard length* (Fig. 1) is the distance to the last vertebra which can be determined as approximately the flexure line or crease caused by bending the caudal fin. This is the measurement usually referred to in this book.

The number of scales on various parts of the body are useful aids in classification. The number is seldom constant but usually fluctuates within a definite range. *The number of scales in the lateral line* (Fig. 2) is an important measurement. Careful counting, often under magnification, is necessary. The pored scales can be counted to the end of the caudal vertebrae which can be determined as for the standard length. These counts usually vary within certain limits for each species. When the lateral line is incomplete or undeveloped, the number of vertical scale rows is commonly substituted. The number of scales in a row between the lateral line and the anterior base of the dorsal fin is designated as the scales above the *lateral line* (Fig. 2). Counts of the scales in a row from the lateral line to front of base of anal fin is known as the *scales below the lateral line* (Fig. 2). The number of mid-dorsal scales anterior to the dorsal fin, and the number of scale rows before the dorsal fin are frequently used in the identification of some species.

General References

Alexander, R.McN. 1967. Functional Design in Fishes. Hutchinson and Co., Ltd., London. 160 pp.

Anderson, D.P. 1974. Diseases of Fishes. Book 4: Immunology. T.F.H. Publ., Neptune, N.J. 239 pp.

Breder, C.M., Jr. and D.E. Rosen. 1966. Modes of Reproduction in Fishes. Natural History Press, Garden City, N.Y. 941 pp.

Brown, M.E. 1957. The Physiology of Fishes. Vol. I. Metabolism. 447 pp. Vol. II. Behavior. 526 pp. Academic Press, New York, N.Y.

Cahn, P.H. 1967. Lateral Line Detectors. Indiana Univ. Press, Bloomington. 496 pp.

Goddrich, E.S. 1958. Studies on the Structure and Development of Vertebrates. Vol. I, Dover Publ. Inc., New York, N.Y. 485 pp.

Gosline, W.A. 1971. Functional Morphology and Classification of Teleostean Fishes. Univ. Press of Hawaii, Honolulu. 208 pp.

Hardisty, M.W. and I.C. Potter. 1971. The Biology of Lampreys. Vol. I. Academic Press, London. 423 pp. 1972. Biology of Lampreys. Vol. II. Academic Press, London. 466 pp.

Hasler, A.D. 1966. Underwater Guideposts. Univ. Wisconsin Press, Madison.

Hoar, W.S. and D.J. Randall. 1969. Fish Physiology. Vol. 2. The Endocrine System. Academic Press, New York, N.Y. 446 pp.

———. 1970. Fish Physiology. Vol. 4. The Nervous System. Academic Press, New York, N.Y. 532 pp.

Ivlev, V.S. 1961. Experimental Ecology of the Feeding of Fishes. Yale Univ. Press, New Haven. 302 pp.

Jones, I.C. 1960. Hormones in Fishes. Symposium Zool. Soc. London No. 1. 181 pp.

Kleerekoper, H. 1969. Olfaction in Fishes. Indiana Univ. Press, Bloomington. 222 pp.

Lagler, K.F., J.E. Bardach and R.R. Miller. 1962. Ichthyology. John Wiley and Sons, Inc., New York, N.Y. 545 pp.

Love, M. 1970. The Chemical Biology of Fishes. Academic Press, New York, N.Y. 547 pp.

Marshall, N.B. 1966. The Life of Fishes. World Publ. Co., New York, N.Y. 402 pp.

———. 1971. Explorations in the Life of Fishes. Harvard Univ. Press, Cambridge, Mass. 204 pp.

Nikolsky, G.V. 1963. The Ecology of Fishes. Academic Press, New York, N.Y. 352 pp.

Norman, J.R. 1975. A History of Fishes. 3rd Ed. revis. by P.H. Greenwood. Halsted Press, John Wiley and Sons, New York, N.Y. 467 pp.

Protasov, V.R. 1970. Vision and Near Orientation of Fish. Israel Sci. Transl., Jerusalem. 174 pp.

Romer, A.S. 1971. The Vertebrate Body. 4th Ed. W.B. Saunders Philadelphia. 452 pp.

Satchell, G.H. 1971. Circulation in Fishes. Cambridge Monographs in Exp. Biol., Cambridge Univ. Press, England. 131 pp.

Scroder, J.H. 1973. Genetics and Mutagenesis of Fish. Springer-Verlag, New York, N.Y. 356 pp.

How to Identify a Fish

Use of the key to the freshwater fishes of the United States and Canada assumes that the reader is familiar with the anatomy and morphology of fishes as discussed in the preceding section of the book. If the family to which the fish belongs is unknown, turn to the key to the families of freshwater fishes (Page 16). To use the key begin with couplet 1a and 1b and select the alternative that accurately describes the structure of the fish. Continue to the couplet indicated by the number at the right-hand side of the key and once again choose between the two alternatives. Proceed through the key in this manner until the family name is found. For each family there is a line drawing with arrows to indicate the diagnostic feature or features used in the key to identify the family. Preceding the family name will be a page number where the key to the genera and species in the family is to be found.

In identifying the species use the same procedure used in identifying the unknown family until the species is found. Many of the species are illustrated and all have a short paragraph following the species common name and scientific name. The latter paragraph will usually give additional characteristics of the species, such as coloration, scale counts, fin ray counts, body measurements, and maximum size. The known distribution of the species will always be given since it can often be of great importance in identification. If the fish you identified was collected from the Missouri River in Iowa but its known distribution is the Atlantic coastal drainage of the southeastern United States, it is quite probable that you misidentified the fish. If this happens you can either start over or simply backtrack through the key until you find the couplet or couplets where you made an incorrect choice. If the specimen in hand does not resemble the illustration, you may have also made a wrong choice and will need to check your identification again.

In those families with a large number of species (Cyprinidae, Ictaluridae, Percidae, etc.) not all the species are illustrated and several species may be listed under the couplet selected. When this happens you should read carefully the descriptive paragraph associated with the name of the species. In this paragraph, characteristics useful in separating the several species are usually listed; and those characters, plus the geographical distribution of the species, may be of great importance in making your final specific determination.

Fish species display considerable morphological variation and it is difficult in a key to encompass all the variation that may be exhibited by the species throughout its range.

Sexual dimorphism is the most obvious variation in fish. Breeding males may have very brilliant colors, breeding tubercles, enlarged fins, etc. As such, they may differ markedly from non-breeding males as well as females. Breeding individuals and immature individuals can be difficult to identify but when such differences are very pronounced there is usually a short phrase in the species description noting such differences. Also when a species is variable it may key out in several couplets and you will be referred to a page or couplet where the species is illustrated and where the descriptive paragraph is located.

Once you have identified the fish you may wish to know more about it than can be included in this book. At the end of the key portion of the book is a selected list of books where more detailed information on the biology and the ecology of the species may be found. In these books you will find very detailed bibliographies where there are citations to the original descriptions of the species.

Key to the Families of Fishes Found in the Fresh Waters of the United States*

1a Mouth without jaws and within a funnel-like depression lined with horny teeth; no paired fins; nostrils consist of an unpaired median pit: seven separate gill apertures on each side. Fig. 13 . . . . **(p. 26) LAMPREY FAMILY,** *Petromyzonidae*

Figure 13.

1b Mouth with upper and lower jaws and not located in a funnel-like depression; nostrils consist of paired openings; one or two pairs of fins present; gills covered by a bony flap or opercle. Fig. 2 2

2a Caudal fin is typical or modified heterocercal type. Figs. 3, 14 3

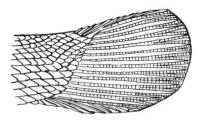

Figure 14.

2b Caudal fin is homocercal type. Fig. 4 . . . 6

3a Caudal fin is typically heterocercal (Fig. 3); mouth is under and behind tip of projecting snout (sub-terminal) and with no or with poorly developed teeth 4

3b Caudal fin is modified heterocercal type (Fig. 14); mouth is located at tip of snout (terminal) and has well-developed teeth 5

4a No scales or bony plates apparent on body; snout is very long and paddle-like; two tiny barbels in front of mouth. Fig. 15 . . . . **(p. 35) PADDLEFISH FAMILY,** *Polyodontidae*

Figure 15.

*Many families of marine fishes contain species which occasionally invade fresh and brackish waters at the mouths of rivers. Some, but not all, of these families are included here.

4b Prominent bony plates on head; 5 rows of keeled plates on body; snout is shovel-like; 4 well-developed barbels in front of mouth. Fig. 16. . . . . (p. 32) **STURGEON FAMILY,** *Acipenseridae*

Figure 16.

5a Jaws very elongate; body covered with hard diamond-shaped or ganoid scales (Fig. 6); dorsal fin short and near caudal fin. Fig. 17 . . . . (p. 36) **GAR FAMILY,** *Lepisosteidae*

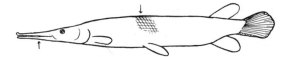

Figure 17.

5b Jaws not elongate; body covered with cycloid scales (Fig. 7); dorsal fin very long, extending over most of the back and almost to the caudal fin. Fig. 18. . . . . (p. 36) **BOWFIN FAMILY,** *Amiidae*

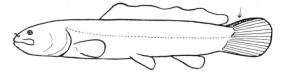

Figure 18.

6a Body elongate or eel-like 7

6b Body not elongate or eel-like 9

7a Snout elongate with small mouth at tip; body covered with annular rings or plates;

dorsal fin small and not reaching caudal fin. Fig. 19 . . . . (p. 150) **PIPEFISH FAMILY,** *Syngnathidae*

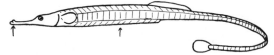

Figure 19.

Marine, but occasionally enter fresh water.

7b Snout not elongate; body covered with scales, scales may be small and difficult to see; dorsal fin long. 8

8a Dorsal fin extending from caudal fin to head, but not continuous with caudal fin. Fig. 20. . . . . (p. 196) **GUNNEL FAMILY,** *Pholidae*

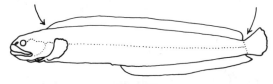

Figure 20.

Marine, but occasionally enter fresh water.

8b Dorsal fin extending from about the middle of the body and continuous with the caudal fin. Fig. 21 . . . . (p. 39) **FRESHWATER EEL FAMILY,** *Anguillidae*

Figure 21.

9a Pelvic fins near anus, abdominal in position. Fig. 22. 10

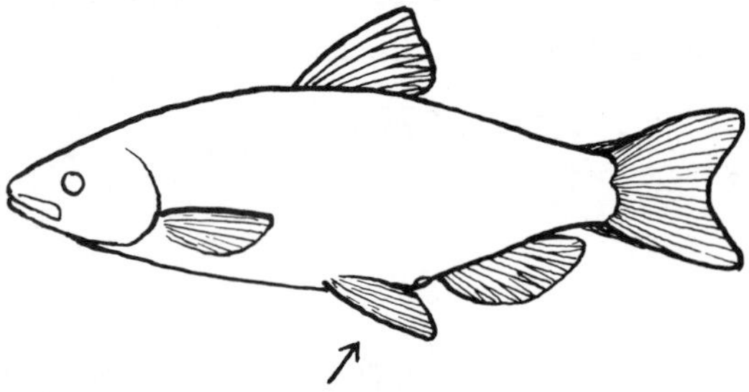

Figure 22.

9b Pelvic fins near, under or in front of pectoral fins, thoracic or jugular in position. Fig. 23. 32

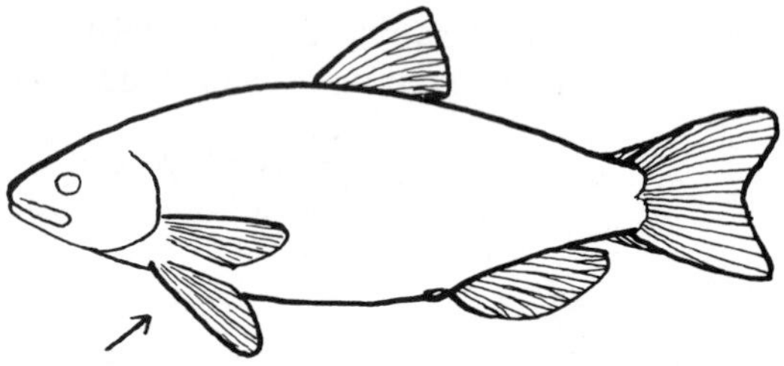

Figure 23.

10a Head covered with bony scutes; with 3 or more bony scutes alongside of body. Fig. 24. . . . . (p. 131) ARMORED CATFISH FAM- ILY, *Loricariidae*

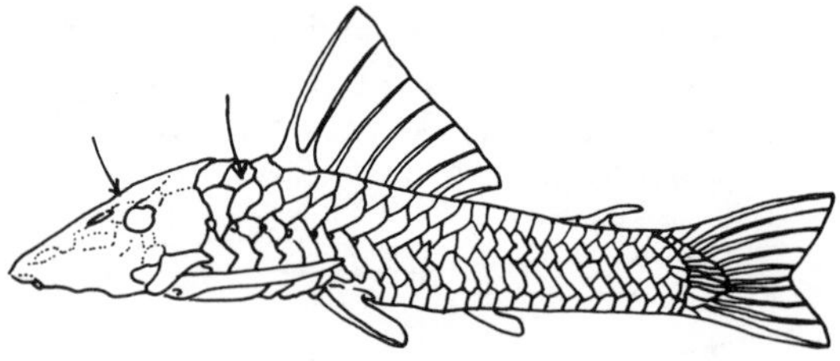

Figure 24.

10b Head without bony scutes; no bony scutes on side of body 11

11a Head without scales. 12

11b Head more or less covered with scales 25

12a Fins usually without spines, spines present only in fins of introduced minnows (carp and goldfish) and some desert minnows from southwestern United States 13

12b Fins with both spines and soft rays. . . . 22

13a Four or more branchiostegal rays (Fig. 9) present on each side 14

13b Less than four branchiostegal rays present on each side 20

14a No adipose fin present 15

14b Adipose fin present. Fig. 2 19

15a Belly with rounded and smooth margin 16

15b Belly with saw-toothed or with a knife-like margin 17

16a Last ray of dorsal fin greatly elongated. Fig. 25. . . . . (p. 38) TARPON FAMILY, *Megalopidae*

Figure 25.

Marine, sometimes enters fresh water.

16b Last ray of dorsal fin not elongated. Fig. 26. . . . . (p. 39) TEN POUNDER FAMILY, *Elopidae*

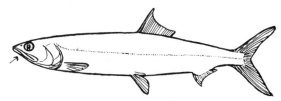

Figure 26.

17a Mouth very large with maxillary extending behind posterior margin of eye. Fig. 27. . . . . (p. 42) ANCHOVY FAMILY, *Engraulidae*

Figure 27.

17b Mouth not large; maxillary does not extend behind posterior margin of eye. 18

18a Lateral line absent; belly with saw-toothed margin entire length. Fig. 28. . . . . (p. 40) HERRING FAMILY, *Clupeidae*

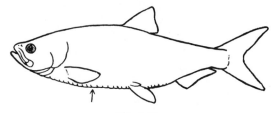

Figure 28.

18b Lateral line present; part of belly with sharp but not saw-toothed margin. Fig. 29. . . . . (p. 42) MOONEYE FAMILY, *Hiodontidae*

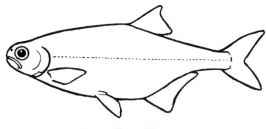

Figure 29.

19a No axillary process present at base of pelvic fin. Fig. 30. . . . . (p. 54) SMELT FAMILY, *Osmeridae*

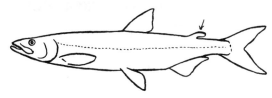

Figure 30.

19b Axillary process present at base of pelvic fin. Fig. 31 . . . . (p. 43) SALMON FAMILY, *Salmonidae*

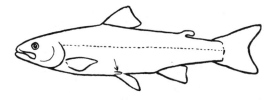

Figure 31.

20a Adipose fin present; teeth are present in mouth. Fig. 32. . . . . (p. 59) CHARACIN FAMILY, *Characidae*

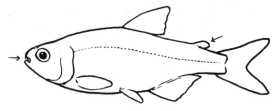

Figure 32.

20b Adipose fin absent; no teeth present in mouth **21**

21a Usually more than 10 dorsal rays: (some western suckers have 9-10 dorsal rays) more than 10 well-developed teeth on each pharyngeal arch confined to a single row (Fig. 33); lips more or less sucker-like, lower lip more or less thick. Fig. 34. . . . . (p. 110) **SUCKER FAMILY,** *Catostomidae*

Figure 33.

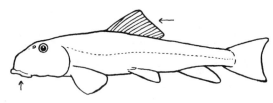

Figure 34.

21b Usually not more than 10 dorsal rays (except in introduced carp, goldfish, and in several western minnows); less than 10 teeth on each pharyngeal arch confined to

2 or 3 rows; or if in a single row, only 4-5 teeth on each side (Fig. 35); lips usually not sucker-like. Fig. 36 . . . . (p. 60) **MINNOW FAMILY,** *Cyprinidae*

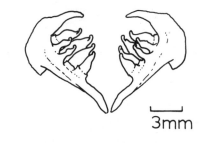

3mm

Figure 35.

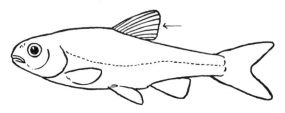

Figure 36.

22a Two to 10 pairs of barbels present; barbels above and below the mouth . . . **23**

22b Barbels lacking or with a single median barbel under the chin. Fig. 49 . . . . **24**

23a Adipose fin present; dorsal fin short, with fewer than 30 soft rays; epibranchial organs absent. Fig. 37 . . . . (p. 121) **FRESHWATER CATFISH FAMILY,** *Ictaluridae*

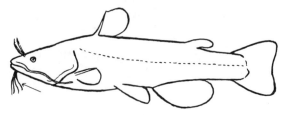

Figure 37.

23b Adipose fin absent; dorsal fin long, with 60 or more soft rays; epibranchial organs highly developed. Fig. 38. . . . . (p. 131) **AIRBREATHING CATFISH FAMILY,** *Clariidae*

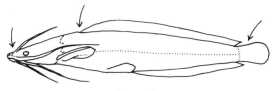

Figure 38.

24a Adipose fin present. Fig. 39. . . . . (p. 134) **TROUTPERCH FAMILY,** *Percopsidae*

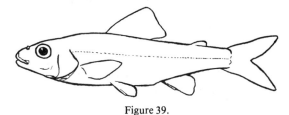

Figure 39.

24b Adipose fin absent. 25

25a Small spiny dorsal fin in front of soft dorsal fin. 26

25b No spiny dorsal fin in front of soft dorsal fin 27

26a Lower jaw more or less extending before upper jaw, snout flattened; eye not partly covered by adipose membrane. Fig. 40 . . . . (p. 147) **SILVERSIDE FAMILY,** *Atherinidae*

Figure 40.

26b Lower jaw does not extend beyond upper jaw; eye partly covered by vertical adipose membrane. Fig. 41 . . . . (p. 193) **MULLET FAMILY,** *Mugilidae*

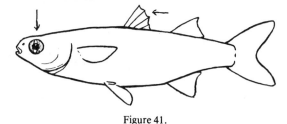

Figure 41.

Marine, sometimes enter fresh water.

27a Lateral line ventral in position; about 300 scales in lateral line; jaws very long and slender. Fig. 42 . . . . (p. 135) **NEEDLEFISH FAMILY,** *Belonidae*

Figure 42.

27b Lateral line imperfect or absent; dorsal in position when present; jaws variable, but not slender; less than 200 scales in lateral line 28

28a Both jaws extend forward and shaped like a duckbill. Fig. 43 . . . . (p. 56) **PIKE FAMILY,** *Esocidae*

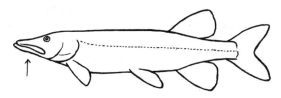

Figure 43.

28b Both jaws not extending far forward and not shaped like a duckbill. 29

29a Lower jaw not protruding. Fig. 44. . . . . (p. 58) MUDMINNOW FAMILY, *Umbridae*

Figure 44.

29b Lower jaw protruding 30

30a Eye degenerate or small; pelvic fins minute or absent; anus tends to be jugular. Fig. 45 . . . . (p. 132) CAVEFISH FAMILY, *Amblyopsidae*

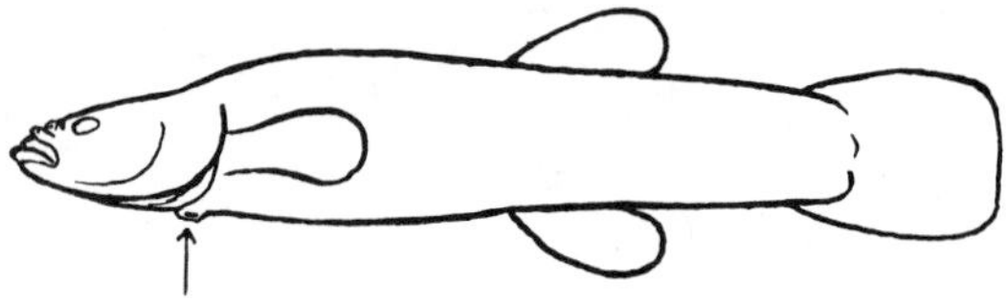

Figure 45.

30b Eyes normal; pelvic fins usually well developed (except in a few desert species) 31

31a Third anal ray (including rudimentary rays) not branched (Fig. 46); anal fin of male modified into elongated intromittent organ (gonopodium). Fig. 47. . . . . (p. 145) TOPMINNOW or LIVE-BEARER FAMILY, *Poeciliidae*

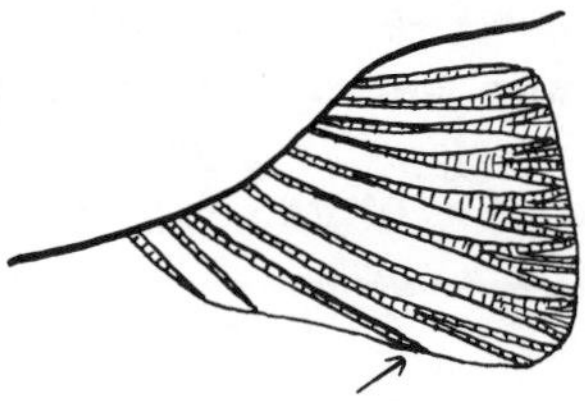

Figure 46.

Figure 47.

31b Third anal ray branched; may not be completely divided in immature individuals; in some the second ray is also branched; anal fin of males not modified. Fig. 48. . . . . (p. 135) KILLIFISH FAMILY, *Cyprinodontidae*

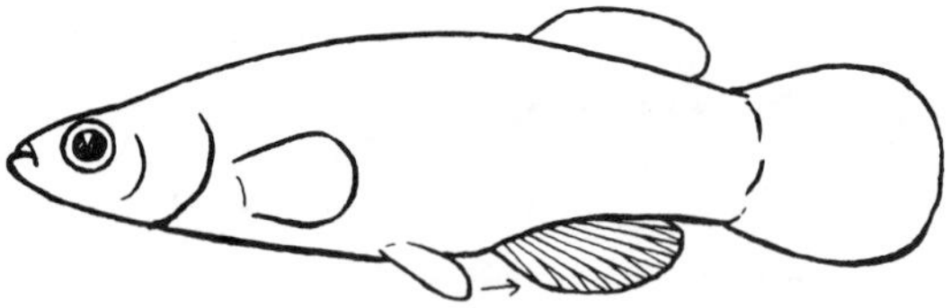

Figure 48.

32a Fins without spines or hard rays. 33

32b Fins with spines or hard rays 36

33a Both eyes on one side of head; body compressed laterally and fish lives on its sides; (FLATFISHES, marine, but some species are frequent invaders of fresh water) 34

33b Eyes normal; body not compressed laterally; median barbel under chin. Fig. 49. . . . . (p. 134) **COD FAMILY,** *Gadidae*

Figure 49.

34a Margin of preopercle hidden by skin; left pectoral fin rudimentary or absent, right pectoral fin may or may not be present. Fig. 50. . . . . (p. 196) **SOLE FAMILY,** *Soleidae*

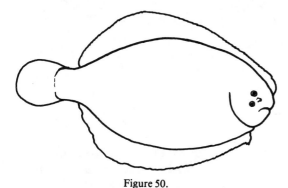

Figure 50.

Marine, sometimes enter fresh water.

34b Margin of preopercle not obscured by skin; both pectoral fins present 35

35a Pelvic fins not symmetrical but pelvic fin on eyed side located on ventral margin. Fig. 51. . . . . (p. 195) **LEFTEYE FLOUNDER FAMILY,** *Bothidae*

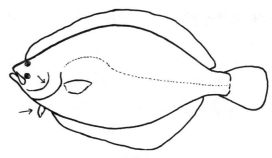

Figure 51.

35b Pelvic fins symmetrical, pelvic fin on eyed side not on ventral margin. Fig. 52. . . . . (p. 195) **RIGHTEYE FLOUNDER FAMILY,** *Pleuronectidae*

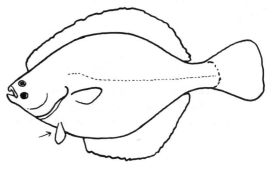

Figure 52.

36a Adults with anus anterior to usual position, usually under throat (jugular). Fig. 53. . . . . (p. 133) **PIRATE PERCH FAMILY,** *Aphredoderidae*

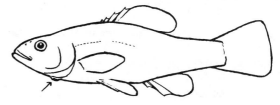

Figure 53.

36b Adults with anus in normal position. . . 37

37a Body without scales, naked or covered with tiny spines or with plates 38

37b Body with scales 39

38a Free spines in front of soft dorsal fin; pelvic fin formed of one spine. Fig. 54 . . . . (p. 148) STICKLEBACK FAMILY, *Gasterosteidae*

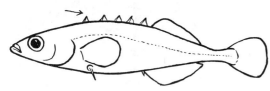

Figure 54.

38b Dorsal spines not free but united to each other by fin membrane; pelvic fins with 3 or 4 soft rays; pectoral fins very large. Fig. 55 . . . . (p. 151) SCULPIN FAMILY, *Cottidae*

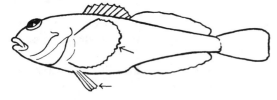

Figure 55.

39a Dorsal fin with 16 or more spines 40

39b Dorsal fin with less than 16 spines 41

40a Distinct ridge of scales along base of dorsal fin; lateral line complete. Fig. 56 . . . . (p. 191) SURFFISH FAMILY, *Embiotocidae*

One freshwater species, one marine species may enter fresh water.

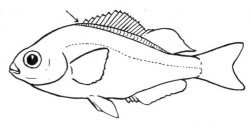

Figure 56.

40b No distinct ridge of scales at base of dorsal fin; lateral line broken under posterior part of dorsal fin. Fig. 57 . . . . (p. 192) CICHLID FAMILY, *Cichlidae*

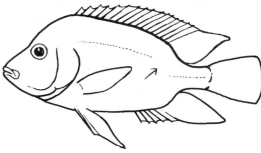

Figure 57.

41a Dorsal fin with 6 to 8 rather filamentous spines; pelvic fins close together, sometimes united. Figs. 58, 59 . . . . (p. 193) GOBY FAMILY, *Gobiidae*

Figure 58.

Mostly marine, several species enter fresh water.

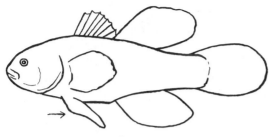

Figure 59.

41b Dorsal fin with 4 to 15 rather stiff spines; pelvic fins never united 42

42a Dorsal fin with 4 spines; pectoral fins on upper half of body. (See Fig. 41.) . . . . (p. 193) MULLET FAMILY, *Mugilidae*

42b Dorsal fin with 6 to 15 spines; pectoral fins on lower half of body 43

43a Anal spines 3 or more 44

43b Anal spines less than 3 45

44a Opercles with well-developed spine; well-developed patch of gill filaments (pseudobranchia) on inner surface of opercle. Fig. 60 . . . . (p. 157) TEMPERATE BASS FAMILY, *Percichthyidae*

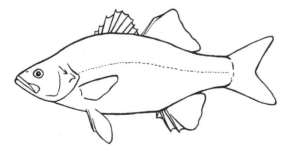

Figure 60.

44b Opercles without a well-developed spine; poorly developed and partly concealed vestigial gill filaments (pseudobranchiae) on the inner surface of the opercle. Fig. 61. . . . . (p. 158) SUNFISH FAMILY, *Centrarchidae*

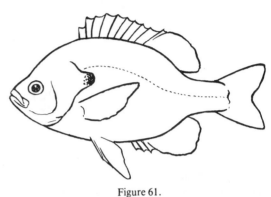

Figure 61.

45a Second anal spine broad and long; lateral line extends onto the caudal fin. Fig. 62. . . . . (p. 190) DRUM or SHEEPHEAD FAMILY, *Sciaenidae*

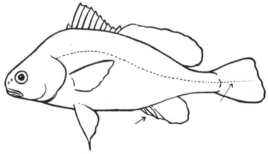

Figure 62.

45b Second anal spine not very broad and long; lateral line not extending onto caudal fin. Fig. 63 . . . . (p. 167) **PERCH FAMILY,** *Percidae*

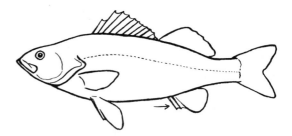

Figure 63.

LAMPREY FAMILY
Petromyzonidae

The members of the lamprey family are eel-like forms with a sucker-disc mouth structure (buccal funnel) filled with horny spines called teeth. The lampreys are highly specialized descendants of the earliest known type of vertebrates and have never possessed upper and lower jaws, true teeth, and paired fins. Their skeleton is very primitive, consisting chiefly of an incomplete cartilaginous brain case and a notochord on top of which are vestiges of vertebrae. No true bone is present. The gills are highly specialized and lie in separate pockets represented by seven clefts on each side of the body. They possess a long dorsal fin which is more or less continuous with the caudal fin. Lampreys vary in size from six inches to several feet. Most of the freshwater species are pale brown or fawn color.

Adult lamprey are modified for a parasitic or predaceous life, possessing a sucker-disc by which they can attach to fishes and rasp a hole for gorging on the blood and tissues. The adults of some species have abandoned this mode of feeding, and do not feed after metamorphosis, but live just long enough in the adult stage to reproduce. The latter, non-predaceous species, have smaller buccal funnels and usually feeble buccal teeth. The buccal teeth are important characters for identification and their general arrangement and terminology is shown in the diagram (Fig. 64). The mouth in the center of the funnel is armed with a set of cutting or rasping plates (*laminae*). In some species the inner lateral teeth may be double or with two points and are termed *bicuspid*. The number of trunk myomeres or muscle segments is another character useful in the identification of lampreys. This count begins with the first myomere posterior to the last gill opening and ends with the myomere whose posterior angle lies partially or entirely above the cloaca or vent. The count is difficult in preserved material but can be facilitated by scraping the skin toward the posterior end, removing the congealed mucous that may be present. Creases in the skin may also make the count more difficult but by flexing the specimen it is usually possible to distinguish the margins of the myomeres from the creases.

The lampreys deposit their eggs in nests made on the bottom of swift streams by pulling back the pebbles. The eggs hatch into larval forms known as *ammocoetes*. These have undeveloped eyes and no sucking disc. The larval forms spend approximately 4 to 5 years in the bottom mud of streams where they feed on organic detritus, eventually metamorphosing into adults with eyes and sucker-discs (buccal funnels).

Several species live in the sea and enter

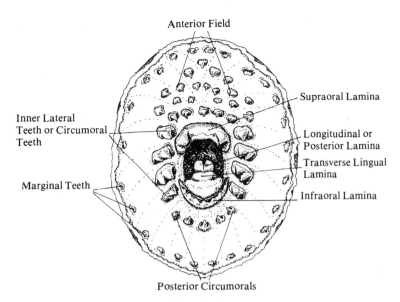

Anterior Field

Supraoral Lamina

Inner Lateral
Teeth or Circumoral
Teeth

Longitudinal or
Posterior Lamina

Transverse Lingual
Lamina

Marginal Teeth

Infraoral Lamina

Posterior Circumorals

Figure 64.

fresh water to spawn. These may become land-locked and develop freshwater races. Other species are entirely freshwater and are found in most of the river systems and lakes of the United States.

1a Dorsal fin with obvious notch, separated into 2 lobes, either separate or joined by a low connection **2**

1b Dorsal fin continuous, not divided into 2 obvious lobes (*Ichthyomyzon*) **10**

2a More than 3 radiating rows of teeth, 4 or more teeth in each row on each side of opening in buccal funnel (Fig. 65); parasitic or predaceous; may reach a length of 2 feet. Fig. 66 . . . .
. . . . **SEA LAMPREY,**
Petromyzon marinus **Linnaeus**

Figure 65.

Figure 66.

Brownish and strongly mottled. Anadromous on the Atlantic coast. The landlocked form, *Petromyzon marinus dorsatus* Wilder is a dwarfed form found in some eastern lakes. The sea lamprey is a recent invader of the Great Lakes other than Ontario where it was native.

2b Scattered groups of teeth not in radiating rows, usually 3 groups of one or two large teeth on each side of opening in buccal funnel 3

3a Upper or supraoral plate with three well-developed teeth (plate with well-developed teeth above opening in the buccal funnel), Fig. 67; 4 pairs of lateral circumoral teeth. (*Entosphenus*) 4

3b Upper or supraoral plate with 2 well-developed teeth (Fig. 74); 3 pairs of lateral circumoral teeth 6

4a Lateral circumoral teeth with cusps typically 2-3-3-2; posterior circumoral teeth usually 14 or more. 5

4b Lateral circumoral teeth with cusps usually 2-2-2-2 or 2(1)-2-2-2(1), very rarely 2-3-3-2; posterior circumoral teeth 13 or less. Figs. 67, 68 . . . .
. . . . **PIT-KLAMATH BROOK LAMPREY,** *Entosphenus lethophagus* (Hubbs)

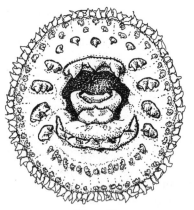

Figure 67.

Figure 68.

A small non-predaceous lamprey, less than 8 inches long. With teeth reduced in size and weak; trunk myomeres 59-65; occasionally there are only 2 well-developed cusps on the supraoral lamina. Klamath River drainage of southern Oregon, Pit River system of Sacramento River, northern California.

5a Trunk myomeres 65 or less. Figs. 69, 70. .
. . . . **MILLER LAKE LAMPREY,**
Entosphenus minimus **(Bond and Kan)**
A small, 6 inches total length, predaceous lamprey. With 13-17 posterior circumorals; trunk myomeres 59-66. Transverse lingual lamina with small well-developed cusps. Restricted to Miller Lake, Oregon and thought to be extinct.

Figure 69.

Figure 70.

KLAMATH RIVER LAMPREY, *Entosphenus folletti* Vladykov and Kott. Fig. 71. A relatively small, 11 inches total length or less total length, non-predaceous lamprey; with 13-18 posterior circumoral teeth; trunk myomeres 61-65. Transverse lingual lamina smooth or with obsolete cusps. Klamath River drainage of northern California.

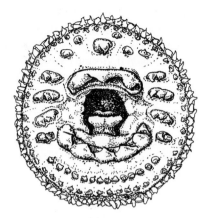

Figure 71.

5b **Trunk myomeres more than 65. Figs. 72, 73** . . . .
. . . . **PACIFIC LAMPREY,** *Entosphenus tridentatus* **(Gairdner)**
Predaceous anadromous lamprey that may reach a length of 2 feet. Pacific coastal streams from southern California northward to the Bering Sea and Unalaska. A few records from Japan but no valid records from mainland Asia.

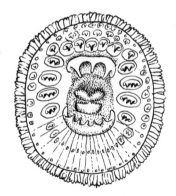

Figure 72.

Figure 73.

6a **Row of small teeth in posterior field of buccal disc between marginal teeth and oral opening. Fig. 74. (*Lethenteron*)** . . . 7

6b **No teeth in posterior field of buccal disc except small marginals. Figs. 76, 78. (*Lampetra*)** 8

7a **Trunk myomeres 50 to 58, average 54** . . . . **GULF BROOK LAMPREY,** *Lethenteron meridionale* **Vladykov, Kott and Pharand-Coad**
Non-predaceous lamprey with dark pigment on second dorsal fin. Small usually less than 6 inches long. Freshwater streams in eastern Gulf coast drainage, Tennessee, Alabama, and Tombigee River systems.

7b **Trunk myomeres 64 to 74, average 68** . . . . . . **AMERICAN BROOK LAMPREY,** *Lethenteron lamottenii* **(Lesueur)**
Figs. 74, 75. Small non-predaceous lamprey common in small streams northward from Missouri, Tennessee to the Great Lakes drainage, including the Saint Lawrence River basin. The Atlantic coastal drainage from North Carolina to Connecticut and New Hampshire; in the Yukon and Mackenzie Rivers, Alaska and Canada, possibly represented by another species. Length usually 6-7 inches, but a giant form 12 inches long has been taken in Lakes Michigan and Huron.

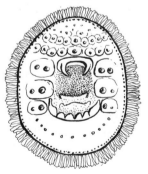

Figure 74.

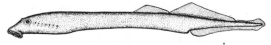

Figure 75.

ARCTIC LAMPREY, *Lethenteron japonicum* (Martens). A large, 12-14, inches long, anadromous predaceous species that is similar to the American brook lamprey. Arctic Ocean drainage of Yukon and Northwest Territories and ranges through Bering Sea drainages of Alaska into Siberia.

8a Trunk myomeres 60 or more 9

8b Trunk myomeres 59 or fewer
(occasionally 60). Figs. 76, 77 . . . .
. . . . LEAST BROOK LAMPREY,
Lampetra aepyptera (Abbott)

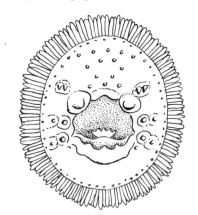

Figure 76.

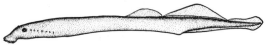

Figure 77.

Small non-predaceous lamprey living in small streams. Trunk myomeres range from 50-59. Upper Ohio River drainage; Atlantic coast from the Potomac to Neuse River; Gulf Coastal Plain in Alabama and Mississippi.

PACIFIC BROOK LAMPREY, *Lampetra pacifica* Vladykov. Small, less than 9 inches long, non-predaceous species. Trunk myomeres 52-58. Pacific coastal streams from San Francisco Bay north to the lower Columbia River in Oregon.

9a Inner lateral teeth 3; unicuspid or bicuspid, or middle tooth bicuspid; teeth poorly developed. Figs. 78, 79 . . . .
. . . . . . WESTERN BROOK LAMPREY,
Lampetra richardsoni Vladykov and Follett

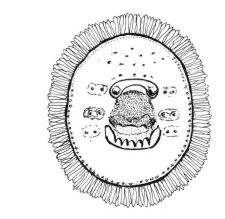

Figure 78.

Figure 79.

Small lamprey with dark blotch at tip of caudal fin; 7-9 inches in length. Coastal streams from Oregon northward to British Columbia.

9b Inner lateral teeth 3; outer pairs bicuspid, center pair tricuspid; teeth hooked and sharp. Figs. 80, 81 . . . .
. . . . . AMERICAN RIVER LAMPREY,
Lampetra ayresi (Günther)

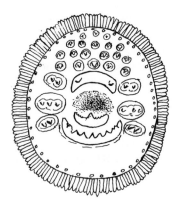

Figure 80.

Figure 82.

Figure 81.

Figure 83.

A predaceous lamprey 5 to 12 inches in length. Anadromous in Pacific coastal streams from central California to British Columbia.

In the upper Mississippi drainage and in parts of Hudson Bay and Great Lakes drainages.

10a Predaceous; diameter of buccal funnel larger than width of head 11

10b Non-predaceous; diameter of buccal funnel not larger than width of head . . 13

11a Teeth on each side of oral opening in buccal funnel (circumoral teeth) mostly unicuspid or with one point (Fig. 82): may reach a length of 15 inches. Fig. 83 . . . . SILVER LAMPREY, *Ichthyomyzon unicuspis* Hubbs and Trautman

11b Teeth on each side of oral opening in buccal funnel (circumoral teeth) mostly bicuspid or with 2 points 12

12a Trunk myomeres 51 to 54; length up to 15 inches . . . . CHESTNUT LAMPREY, *Ichthyomyzon castaneus* Girard

Resembles the silver lamprey but differs in having well-developed bicuspid teeth on each side of buccal funnel (Fig. 84). Northern Wisconsin

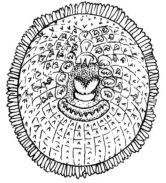

Figure 84.

to Louisiana and Alabama, also in western Manitoba and in parts of Great Lakes drainage.

12b Trunk myomeres usually 56 to 58; length up to 12 inches. (See Fig. 82.) ....
.... **OHIO RIVER LAMPREY,**
Ichthyomyzon bdellium, **Jordan**

Resembles silver lamprey. Ohio River system.

13a Teeth small and poorly developed; circumoral teeth not bicuspid (Fig. 85); may reach a length of 10 inches. Fig. 86. .
.... **NORTHERN BROOK LAMPREY,**
Ichthyomyzon fossor, **Reighard and Cummins**

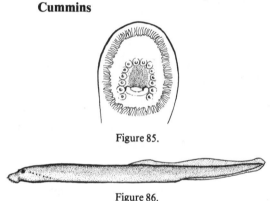

Figure 85.

Figure 86.

Wisconsin eastward in the Great Lakes drainage.

13b Teeth moderately to well developed; circumoral teeth bicuspid 14

14a Trunk myomeres 51 to 54; may reach a length of 9 inches. ....
.... **SOUTHERN BROOK LAMPREY,**
Ichthyomyzon gagei **Hubbs and Trautman**

Resembles northern brook lamprey. Lower Mississippi drainage and Gulf coast drainage from eastern Texas to western Florida.

14b Trunk myomeres 55 to 61; may reach a length of 12 inches. ....
... **ALLEGHENY BROOK LAMPREY,**
Ichthyomyzon greeleyi **Hubbs and Trautman**

Resembles northern brook lamprey. Upper Ohio River drainage. A closely related species, the mountain brook lamprey, *Ichthyomyzon hubbsi* Raney, in the upper Tennessee River drainage, closely resembles the Allegheny brook lamprey but differs in having weaker teeth and a much lower first dorsal fin.

STURGEON FAMILY
Acipenseridae

The sturgeons are modern relicts of some of the early bony fishes. Many of their structures are quite primitive, indicating that they are survivors of an ancient group. They possess a primitive heterocercal caudal fin and a spiral valve intestine. Their skeleton is largely cartilaginous and they retain a notochord. Bony plates cover their heads and extend in several longitudinal rows along their bodies. Scales are mostly restricted to a patch of ganoid scales on the caudal fin.

Sturgeons possess a more or less prolonged shovel-shaped snout under which is a sucker-like mouth with thick lips. The mouth is well adapted for working over the bottom where they pick up small animals for food. A row of sensory barbels before the mouth aids in locating their food.

Sturgeons occur in northern Europe, Asia, and North America. Some attain huge size, some in Russia having been reported as weighing over a ton. Many species are anadromous, living in the sea and entering fresh water to spawn. The three species found in the central United States are strictly freshwater, but those found on the west and east coasts are anadromous.

Sturgeons spawn in the spring, passing upstream to gravel beds where they deposit their eggs. They give no care to the eggs or young. Sturgeons are quite important as the flesh is excellent for food and the eggs are used for caviar.

1a Small opening (spiracle) between eye and upper corner of opercle (Fig. 87); caudal peduncle heavy and not entirely covered by bony plates; lower lip with 2 slightly papillose lobes, none on upper lip (Fig. 88). (*Acipenser*) 3

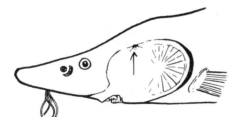

Figure 87.

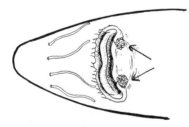

Figure 88.

1b No opening (spiracle) between eye and upper corner of opercle (Fig. 89); caudal peduncle very slender and completely enclosed by plates; lower lip with 4 papillose lobes (Fig. 90). (*Scaphirhynchus*) 2

Figure 89.

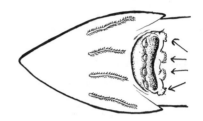

Figure 90.

2a Belly covered with small bony scale-like plates. Fig. 91 . . . . SHOVELNOSE STURGEON, *Scaphirhynchus platorynchus* (Rafinesque)

Figure 91.

Pale brown. Length about 2 feet, although records of 5 feet are reported. Mississippi River and larger tributaries.

2b Belly mostly naked . . . . PALLID STURGEON, *Scaphirhynchus albus* (Forbes and Richardson)

Light brown. Length about 2 feet, although may reach a length of nearly 5 feet. Upper Mississippi River and larger tributaries.

3a Bony plates between pelvic and anal fins in 2 rows of 4 to 8 each (Fig. 92), dorsal rays about 45. Fig. 93 ....
.... **WHITE STURGEON,**
Acipenser transmontanus **Richardson**

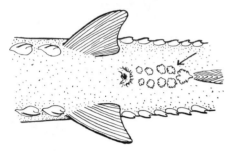

Figure 92.

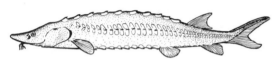

Figure 93.

Grayish brown. Reaches large size of over 8 feet. Anadromous, Pacific coast from Monterey northward into Alaska.

3b Bony plates between pelvic and anal fins in one row of 1 to 4 plates (Fig. 94); dorsal rays less than 45. 4

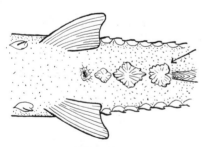

Figure 94.

4a Space between dorsal and lateral rows of plates containing 4 to 10 rows of smaller star-shaped plates 5

4b Space between dorsal and lateral rows of plates containing many rows of minute plates of spicules 6

5a Anal Fin almost as long as dorsal fin and about entirely behind dorsal fin; dorsal rays 33; anal rays 22; about 9 dorsal plates and 26 lateral plates. Fig. 95 ....
.... **GREEN STURGEON,**
Acipenser medirostris **Ayres**

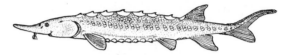

Figure 95.

Greenish color. Size small. Snout more pointed than that of other Pacific species, *A. transmontanus*. Anadromous, Pacific coast from San Francisco northward into Alaska.

5b Anal fin not much more than 1/2 as long as dorsal fin and almost entirely below it; dorsal rays 38; anal rays 27; about 10 dorsal plates and 29 lateral plates. Fig. 96 ...
.... **ATLANTIC STURGEON,**
Acipenser oxyrhynchus **Mitchill**

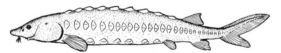

Figure 96.

Olive gray or brown reaching a length of 12 feet. Anadromous, Northern Gulf coast and Atlantic coast into Canada.

6a Front of anal fin below front of dorsal fin and 1/2 as long as dorsal fin; dorsal rays about 41; anal rays about 22; about 8-11 dorsal plates and 22-33 lateral plates. Fig. 97. ....
.... **SHORTNOSE STURGEON,**
Acipenser brevirostrum **Lesueur**

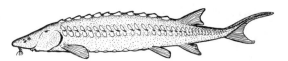

Figure 97.

Length not much over 2 feet, Anadromous, Atlantic coast from Florida to Cape Cod.

6b Front of anal fin below middle of dorsal fin and about 2/3 as long as dorsal fin; dorsal rays 35-39; and rays 22-28; about 15 dorsal plates and 30 to 38 lateral plates.

Fig. 98. . . . .
. . . . LAKE STURGEON,
Acipenser fulvescens **Rafinesque**

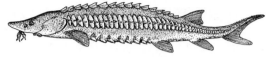

Figure 98.

Brownish sometimes mottled. Lengths of over 7 feet are known. Freshwater, Great Lakes, and Upper Mississippi drainages, and in the Saskatchewan and Hudson Bay drainages.

PADDLEFISH FAMILY
Polyodontidae

The paddlefish family contains only one American species, the paddlefish, *Polyodon spathula* (Walbaum) (Fig. 99) found in the larger streams and connected waters of the Mississippi drainage. Another species, *Psephurus gladius* is found in China. These represent relics of an ancient and primitive group. They possess a heterocercal caudal fin and a spiral valve intestine. The internal skeleton is mostly cartilage.

Figure 99.

The paddlefish is characterized by a long flat snout resembling a paddle. The body is covered by smooth skin and the only evidence of scales is a small patch of ganoid scales on the caudal fin. The gills are covered by opercles with long pointed flaps reaching far back on the body. The gill rakers are filamentous and form a very efficient sieve for straining out the food. These fishes swim about with their mouths open, allowing the water to pass in and out through the gills straining out the plankton crustacea and other small animals on which they feed. They reach a length of over eight feet and a weight of over 200 pounds.

Until recently little was known about the reproductive habits of the paddlefish. They spawn in the spring over gravel bars in swift water. The young are hatched without a paddle-snout, and this structure develops as the fish grows. They are an excellent food fish, and their eggs are sometimes used for caviar.

BOWFIN FAMILY
Amiidae

The bowfin family contains but one living species known as the bowfin or freshwater dogfish, *Amia calva* Linnaeus (Fig. 100), which is generally recognized as a survivor of an early primitive group. They retain much cartilage in their skeleton and have a sheath of bony plates covering their semi-cartilaginous skull. Under the throat a bony plate, known as the gular plate, fills the space between the lower jaws. The young are hatched with a heterocercal caudal fin which changes with growth into a modified heterocercal type. The body is covered by cycloid scales.

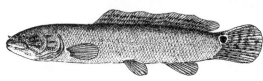

Figure 100.

This fish is readily identified by its long dorsal fin reaching almost to the caudal fin. Bowfins are olive green on the back, shading lighter on the sides to yellow on the belly. Their backs and sides are more or less mottled. The males have an ocellus or "eye spot" at the base of the caudal fin. The lower fins become a vivid blue-green during the breeding season. Bowfins retain a connection between the air bladder and the pharynx which enables them to use the air bladder as a respiratory organ. They rise frequently to the surface and take a fresh "breath" which enables them to live in stagnant waters where the oxygen may be insufficient for most other fishes.

Bowfins spawn in the spring. The male clears out a bowl-like depression in shallow water and guards the eggs and newly hatched fry for several weeks. Bowfins feed on all sorts of living animals, preying heavily on small fishes. Their flesh is not very palatable, and they are seldom used for food. They reach a length of over 2 feet and a weight of ten pounds.

Bowfins occur in sluggish rivers and shallow lakes of the Mississippi drainage from Minnesota eastward through part of the Great Lakes and St. Lawrence drainage and south to the Gulf. They range southward in the Atlantic drainage from Connecticut to Florida.

GAR FAMILY
Lepisosteidae

The gar family contains about 7 species which are restricted to Central and North America and the West Indies. The gars are primitive fishes, retaining much cartilage in their skeleton. Their heads are covered with bony plates. They are characterized by a long cylindrical body covered with ganoid scales (Fig. 6) and by long jaws heavily armed with sharp teeth.

Six species of gars are found in the United States east of the Rockies. These inhabit warm sluggish waters where they lie in wait for their prey. They feed on all kinds of fishes both dead and alive. They have an air bladder which retains a wide passage to the pharynx, and they use this organ for part of their respiration. Hence, they can live in very stagnant waters. They spawn in the spring depositing their eggs at random in shallow water, giving no care to the eggs or young. The eggs are quite toxic and cause great distress if eaten by warm-blooded vertebrates.

1a **Total gill rakers on left outside arch over 50. Distance from tip of snout to angle of jaw slightly shorter than rest of head (Fig. 101); mouth of adults with 2 rows of large teeth in upper jaw; snout broad and blunt. Fig. 102. . . . .**
. . . . **ALLIGATOR GAR,**
Lepisosteus spatula **Lacépède**

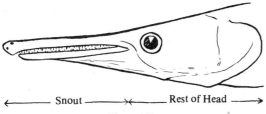

←———— Snout ————→←———— Rest of Head ————→

Figure 101.

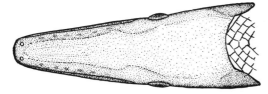

Figure 102.

Greenish or olivaceous above, pale below; snout broader than in shortnose gar. Reaches a very large size of over 12 feet in length. Gulf drainage and Mississippi River north to St. Louis.

1b **Total gill rakers on left outside arch less than 50. Distance from tip of snout to angle of jaw as long or longer than rest of head. Mouth with 1 row of large teeth (teeth of inner row mostly small) in upper jaw 2**

2a **Snout long and slender, length more than twice the distance from angle of mouth to posterior edge of opercle. Fig. 103. . . . . .**
. . . . **LONGNOSE GAR,**
Lepisosteus osseus **(Linnaeus)**

Figure 103.

Olivaceous above and silvery below; a few large spots on body posteriorly. Length up to 5 feet. Southern Minnesota to Vermont and south to Gulf and Rio Grande River.

2b **Snout short and broad. Length less than twice the distance from the angle of the mouth to the posterior edge of the opercle. Fig. 104. . . . . 3**

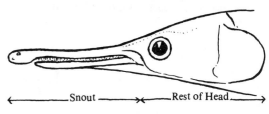

←———— Snout ————→←———— Rest of Head ————→

Figure 104.

3a **Top of head with large round spots; diffuse spots on fins; 54 to 58 scale rows alongside of body. Figs. 105, 106. . . . . . .**
. . . . **SPOTTED GAR,**
Lepisosteus oculatus **(Winchell)**

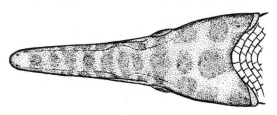

Figure 105.

Figure 106.

Figure 107.

Olivaceous above and silvery below; large spots scattered on body posteriorly. Length up to 4 feet. Iowa and Nebraska to Gulf. A closely allied species, the Florida gar, *Lepisosteus platyrhincus* DeKay replaces this form in Florida and southern Georgia. It differs in having a broader and shorter snout and lacks body plates on isthmus. Length up to 30 inches.

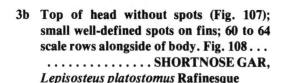

Figure 108.

3b Top of head without spots (Fig. 107); small well-defined spots on fins; 60 to 64 scale rows alongside of body. Fig. 108 . . . . SHORTNOSE GAR, *Lepisosteus platostomus* **Rafinesque**

Olivaceous above and silvery below; large spots on body posteriorly. Length of 3-4 feet. Great Lakes and Mississippi Valley to northeastern Texas.

TARPON FAMILY
Megalopidae

This is a marine family closely related (sometimes included) to the Elopidae, but differing in the lack of pseudobranchiae and in several internal structures. This family represented by the tarpon, *Megalops atlantica* Valenciennes, which is one of the best known game fishes of the southern Atlantic and Gulf of Mexico. The tarpon (Fig. 109) often enters freshwater streams of Florida and other Gulf states. It has a gular plate and is characterized by large silverish scales and a long filamentous posterior ray in the dorsal fin. It reaches a length of over 6 feet but those caught in fresh water are usually much smaller.

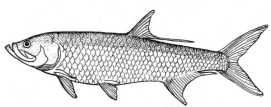

Figure 109.

TEN-POUNDER FAMILY
Elopidae

The members of this family live in the sea, but several species may invade a short distance into fresh water. The ten pounder or big eye herring, *Elops saurus* Linnaeus (Fig. 110) may enter rivers in the Gulf states, and a similar species *Elops affinis* Regan may also enter fresh water in California. These fish are silvery and possess a gular plate between the lower jaws resembling that in *Amia calva*. The eyes are covered by adipose eyelids. They possess pseudobranchiae and long slender gill rakers. The dorsal and anal fins are depressible into a sheath of scales. The ten pounder reaches a length of 3 feet.

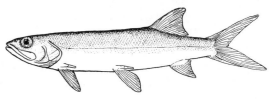

Figure 110.

FRESHWATER EEL FAMILY
Anguillidae

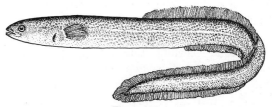

Figure 111.

The freshwater eel family is represented in the United States by one species, the American eel, *Anguilla rostrata* (Lesueur) (Fig. 111). This fish occurs in brackish water along both the Atlantic and Gulf coasts and enters the rivers often penetrating to the headwaters.

The body is very long and slender, reaching a length of over two feet. The skin is yellowish brown and has a smooth appearance as the scales are minute and imbedded. The dorsal fin is very long and is continuous with the caudal and the anal fins. The pelvic fins are absent. The opercular margin is partly fused to the body, leaving only a small gill aperture.

The eel is a catadromous fish, living in fresh water but spawning in the deep Atlantic near Bermuda. The adults apparently die after spawning as they are seen no more. The eels have a very high reproductive potential, females often containing over 10,000,000 eggs. The tiny larval eels after hatching do not look like eels and are called *leptocephala*. They make their way to the mouths of the rivers where they grow into tiny eels four to six inches long. The males remain in the lower part of the river and never become very large. The females make their way upstream to the headwater pools and quiet stretches, traveling mostly at night. The eels remain in fresh water until they are sexually mature, the females reaching a length of 3 to 4 feet in five to seven years. When sexually mature, the females migrate downstream joining the males and swimming out into the ocean.

Eels occur in most of the rivers of the Atlantic and Gulf drainages. Eels are omnivorous, feeding on all kinds of animal food, both dead and alive. They are nocturnal in habits and have the ability to wriggle about on land for several hours. Eels are important for food in many places and are sometimes caught and sold commercially.

HERRING FAMILY
Clupeidae

The herring family contains many important marine species including the true herring. Several species are common in fresh water, including some anadromous species living in the sea, but entering fresh water to spawn.

Members of this family are characterized by having a saw-toothed edge on the belly. They are thin fishes with silvery scales and bluish backs. Many have one or more spots on sides. Some of the anadromous species have spectacular spawning runs in the spring as they crowd upstream to spawn. No care is given to the eggs or young. They feed on a variety of small animal life. Some, such as the gizzard shad, strain out and utilize the larger plankton crustacea. Many species are utilized for food, and all are large valuable forage fishes.

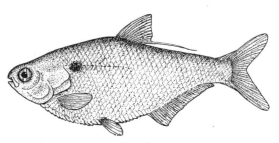

Figure 112.

Length up to 18 inches. Commonly found in fresh water, but may enter brackish water. Minnesota to St. Lawrence River and New Jersey, south to Gulf and into Mexico.

1a Last ray of dorsal fin greatly elongated, forming a long filament. 2

1b Last ray of dorsal fin not elongated 3

2a Anal fin with 30 to 33 rays. Fig. 112. GIZZARD SHAD, *Dorosoma cepedianum* (Lesueur)

2b Anal fin with 20 to 25 rays. . . . . THREADFIN SHAD, *Dorosoma petenense* (Gunther)
Similar to gizzard shad. Length 8-10 inches. Gulf of Mexico, entering streams from Florida into Mexico and penetrating far upstream. Introduced in southwestern U.S. and in Hawaii.

3a Teeth present on rim (premaxillaries) of upper jaw; gill rakers short. Fig. 113 . . . . SKIPJACK HERRING, *Alosa chrysochloris* (Rafinesque)

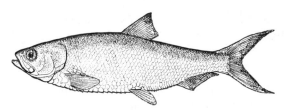

Figure 113.

Bluish above and silvery below. Length up to 15 inches. Gulf of Mexico, entering various river systems including the Mississippi River where it extends upstream to Minnesota.

OHIO SHAD, *Alosa ohiensis* Evermann. Resembles river herring. Lower jaw about equal to length of upper which bears a central notch. Formerly in the Ohio River and parts of the Mississippi River, but has been considered extinct. However, it has been reported from Oklahoma by Dr. George Moore.

3b Teeth more or less restricted to tongue and vomer, absent on rim (premaxillaries) of upper jaw; gill rakers long 4

4a Silvery patch on cheek deeper than long; more than 55 gill rakers on lower part of first gill arch. Fig. 114 . . . .
. . . . AMERICAN SHAD,
***Alosa sapidissima* (Wilson)**

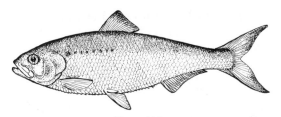

Figure 114.

Bluish above and silvery white below; one or more spots in a longitudinal row behind opercle. Length up to 30 inches. Anadromous, Atlantic coast, entering rivers to spawn. Introduced and now abundant on Pacific coast from California into southern Alaska.

ALABAMA SHAD, *Alosa alabamae* Jordan and Evermann. Resembles the American shad but has only about 40 gill rakers on the lower arm of 1st arch instead of about 60. Anadromous in the Gulf from the Suwanne River to the Mississippi.

4b Silvery patch on cheek longer than deep; less than 55 gill rakers on lower part of first gill arch. Fig. 115 . . . .
. . . . . ALEWIFE, *Alosa pseudoharengus* (Wilson)

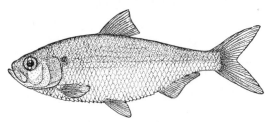

Figure 115.

Bluish above and silvery below; a dark spot behind opercle. Length 15 inches. Anadromous along Atlantic coast and often landlocked. Recently entered the Great Lakes.

HICKORY SHAD, *Alosa mediocris* (Mitchill). Similar to alewife but with row of 5-6 spots behind opercle. Atlantic Ocean from Cape Cod to Florida, sometimes entering river mouths.

BLUEBACK HERRING, *Alosa aestivalis* (Mitchill). Very similar to the alewife, but is more slender and has a dark peritoneum. Anadromous, ranging from Bay of Fundy to Florida. Length 12 inches.

Several species of menhaden, similar to the American shad, but readily recognized by their fluted scales, are common along the Atlantic and Gulf coasts and may enter mouths of rivers.

ANCHOVY FAMILY
Engraulidae

The anchovy family contains small elongated fishes with rather compressed bodies, found in the warmer seas. They are closely allied to the herring family. Several species may enter fresh water.

BAY ANCHOVY, *Anchoa mitchilli* (Valenciennes) commonly invades fresh waters from Massachusetts to Texas. This is a small fish 2 to 4 inches in length with a pale silvery body peppered with tiny black spots and with a faint lateral band. This fish (Fig. 116) is easily distinguished by its large mouth and long maxillary which extends far back of the posterior margin of the orbit. The striped anchovy, *Anchoa hepsetus* (Linnaeus) (Fig. 117) with fewer anal rays (18 to 23) is common along the coast from Cape Cod south and may enter fresh water. It reaches a length of 6 inches.

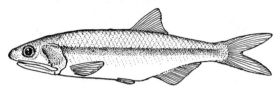

Figure 117.

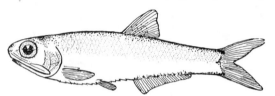

Figure 116.

MOONEYE FAMILY
Hiodontidae

The family Hiodontidae contains only several species, all of which are strictly freshwater. They are thin-bodied, silvery fishes resembling the herrings, but lacking the saw-toothed margin on the belly. The heads are small, and the eyes are large.

They feed on small aquatic organisms, including small fishes. They are utilized to a limited extent for food, mostly as smoked goldeye in the north.

1a Belly keeled anterior to pelvic fins; front margins of anal fin about under or even with front margin of dorsal fin; 9 dorsal

rays usually present. Fig. 118. . . . .
. . . . GOLDEYE, *Hiodon alosoides* (Rafinesque)

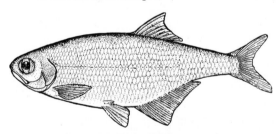

Figure 118.

Silvery somewhat darker on back. Length over 15 inches. Northwestern Canada and Hudson

Bay drainage south to Ohio drainage in Tennessee.

1b   **Belly keeled between pelvic and anal fins; front margin of anal fin is below center of dorsal fin; 11 to 12 dorsal rays present. Fig. 119. . . . . MOONEYE,** *Hiodon tergisus* **Lesueur**

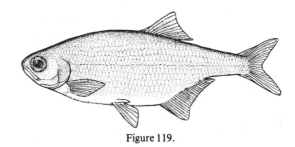

Figure 119.

Silvery with somewhat dusky shades along back. Length 15 inches. Northwest Canada to St. Lawrence and Lake Champlain drainage.

SALMON FAMILY
Salmonidae

The salmon family contains the salmon, trout, whitefishes, and grayling. These fishes are characterized by adipose fins and by an axillary process at the base of each pelvic fin. The family is divided into the salmon-trout sub-family, Salmoninae, the whitefish sub-family, Coregoninae, and the grayling sub-family, Thymallinae, which have often been considered as separate families.

The members of the salmon-trout sub-family, Salmoninae are fine scaled fishes, possessing well-developed teeth and coarse stubby gill rakers. This sub-family includes the Pacific salmon (*Oncorhynchus*), the trout (*Salmo*) which includes the Atlantic salmon, and the charrs (*Salvelinus*) which includes the brook trout, the Dolly Varden trout, the lake trout, and the arctic charr.

These fishes were originally confined to the colder waters of the Northern Hemisphere, but through artificial propagation some have now been distributed to many parts of the world. All are either popular game fishes or important commercial food fishes. They are all cold water fishes, thriving best in water not warmer than 70 °F. Both the Pacific salmon and the Atlantic salmon are anadromous, living in the sea, but spawning in fresh water. The Pacific salmon die after spawning but the Atlantic salmon live to spawn again. Many species of trout develop sea going races. Most trout live in both streams and lakes, except the lake trout which prefers lakes.

Trout are predaceous, feeding on a wide range of small animals, the larger individuals tending to become piscivorous. Most of the species of this family spawn in the fall, usually depositing their eggs on gravel beds. The eggs are covered and left to develop over winter, hatching in the early spring. Spring spawning races are known for several species.

The members of the whitefish sub-family, Coregoninae, are found only in the northern part of the Northern Hemisphere. They have large scales, weak jaws, and many have filamentous gill rakers. Most of them live in lakes, and with the exception of the western and arctic forms seldom enter streams.

They prefer cold water and are particularly abundant in the Great Lakes and other deep northern lakes. A number of species are restricted to the Great Lakes where they are well adapted to live in the deeper water far from shore, feeding on plankton and the deep bottom organisms.

Most of the whitefish spawn in the fall, and the eggs develop during the winter, hatching in early spring. Although a few are caught by hook and line, most of them are caught in nets. Certain species support important commercial fisheries in Canada. Up until the early 1950s several species of whitefish found in the Great Lakes supported a profitable and important fishery. These excellent and once numerous food fishes have been so reduced in numbers that only in Lake Huron and Lake Superior are they fished commercially. The decline in the fishery is mainly the result of the introduction of species such as the lamprey, smelt, and the alewife. The lamprey preyed on the larger species of whitefish once the lake trout had been depleted or eliminated, and the smelt and alewife compete with the various whitefishes for food. This is especially true of the alewife. In the lower Great Lakes a decline in water quality as a result of domestic and industrial pollution has also had an impact on the fishery. Attempts are being made to reduce pollution, the lamprey populations are being controlled, lake trout are being stocked and other salmonids are being introduced to maintain or establish a sport fishery. The smelt is now an important commercial fish but at present there is little that can be done to control the alewife. Once again it is a case of "too little too late." A number of the coregonines once found in the Great Lakes are now extinct and we know them only from preserved specimens in museums or from descriptions in scientific papers. Many that have survived are now endangered. For example, of eight species of *Coregonus* in the Great Lakes Thomas N. Todd lists two as extinct, two as rare or endangered, one as surviving only in Lake Superior, and the remaining three as endangered and all have declined in numbers. It remains to be seen whether our efforts to save these few remaining species will be successful.

Many of the species display considerable variability and it is difficult to ascertain whether the variability is genetic or environmental in origin or, what is more probable, a combination of the two factors. This plasticity is particularly apparent in the cisco, *Coregonus artedi,* where populations from adjacent lakes may be strikingly different one from another. Many of these populations were originally described as separate species, subsequently reduced to sub-species, and finally not recognized taxonomically but considered to be environmentally induced variants. Such variability confounds fish systematists but it is now thought that the variability is merely a reflection of the interactions between these extremely plastic species and their variable environments. When individuals from lakes containing "dwarf" ciscoes are transplanted to nearby lakes they may grow rapidly, doubling in length and tripling in weight in a single year. Variability is the rule in fishes but such tremendous variability is very unusual.

The graylings are represented by several species found in northern Europe, Asia, and North America. They are similar to the trout, but readily recognized by the large sail-like dorsal fin. They spawn in the spring.

The graylings have a beautiful irridescent body colored purplish gray and silver with small spots, and have rows of bluish spots edged with rose or orange on the dorsal fin.

1a   Dorsal fin as long or longer than the head; more than 15 dorsal rays, Fig. 120 . . . . . .
. . . . **AMERICAN GRAYLING,**
Thymallus arcticus **(Pallas)**

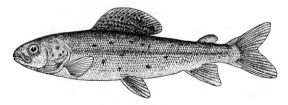

Figure 120.

This is the only species found in North America, and is found in the cold streams of arctic America from Hudson Bay westward. In the United States isolated populations are found in several streams of the Lower Peninsula and the Otter River of the Upper Peninsula of Michigan, and in the headwaters of the Missouri River. They have been introduced elsewhere.

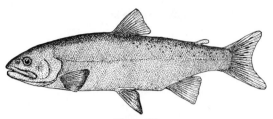

Figure 121.

Silvery fishes with bluish backs, caudal and adipose fin covered with black spots, those on caudal fin are coarse oblong spots. Weight up to 6 lbs. Anadromous, San Francisco to Alaska. Introduced recently into Lake Superior and into the Atlantic Ocean.

1b Dorsal fin shorter than head; less than 15 dorsal rays. 2

2a More than 100 scales in the lateral line; maxillary extends behind center of eye (SALMON-TROUT). 3

2b Less than 100 scales in the lateral line; maxillary does not extend behind center of eye (WHITEFISHES) 16

3a Anal fin with 13 to 19 rays; branchiostegals 13 to 19; vomer narrow and long with weak teeth; body and caudal fin with spots; dorsal fin seldom with spots. . . . . (PACIFIC SALMON, *Oncorhynchus*) 4

3b Anal fin with 7 to 12 rays; branchiostegals usually 10 to 12; body and caudal fin with or without spots; dorsal fin with spots 8

4a Lateral line scales numerous, more than 200. Fig. 121 . . . . PINK SALMON, *Oncorhynchus gorbuscha* (Walbaum)

4b Lateral line scales less than 160 5

5a Gill rakers short, 19-28 on first arch. . . . 6

5b Gill rakers long, 30 to 50 on first arch. Fig. 122 . . . . SOCKEYE SALMON, *Oncorhynchus nerka* (Walbaum)

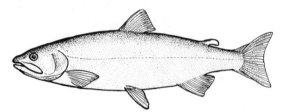

Figure 122.

Bluish above and silvery below; no spots. Length about 2 ft. and weight up to 8 lbs. Anadromous, California to Alaska. The little redfish or kokanee, *O. nerka kennerlyi* (Suckley) is a dwarfed landlocked form. Introduced elsewhere.

6a Anal rays usually 13-15 7

6b Anal rays usually 15-17. Fig. 123 . . . .
. . . . **CHINOOK SALMON,**
Oncorhynchus tschawytscha **(Walbaum)**

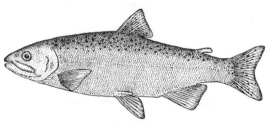

Figure 123.

Dusky or bluish above, silvery below. Flesh red. Back, dorsal and caudal fins usually profusely covered with spots. Weight up to 100 pounds and length of 5 feet. Anadromous, San Francisco to Alaska. Introduced into Lakes Superior and Michigan.

7a Scales above lateral line usually 19-26; scales below lateral line usually 15-24. . . .
. . . . **CHUM SALMON,**
Oncorhynchus keta **(Walbaum)**
Dusky above and light below. Fins more or less blackish. Flesh pale. Weight up to 12 pounds. Anadromous, San Francisco to Alaska.

7b Scales above lateral line usually 23-31; scales below lateral line usually 23-34. Fig. 124 . . . .
. . . . **COHO SALMON,**
Oncorhynchus kisutch **(Walbaum)**

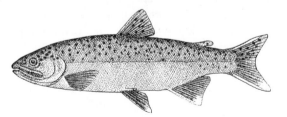

Figure 124.

Bluish above, silvery below. Few spots on back, dorsal fin, and base of caudal fin. Small, length about 2 feet and weight 8 pounds or slightly more. Anadromous, Monterey to Alaska. Introduced recently into the Great Lakes and into the Atlantic Ocean.

8a Body with dark spots on light background; teeth on shaft of vomer arranged in two rows in alternating series or zigzag pattern; scales conspicuous, fewer than 180 in the lateral line (*Salmo*) 9

8b Body with light spots on dark background; teeth on anterior part or head of vomer, not on shaft; scales not conspicuous, more than 190 rows at the lateral line . . . . . . (*Salvelinus*) 13

9a Hyoid teeth (small teeth behind those on tip of tongue) always present; lateral line scales more than 150; red or pink streak on underside of each mandible; dorsal rays 9-11, usually 10. Fig. 125 . . . .
. . . . **CUTTHROAT TROUT,**
Salmo clarki **Richardson**

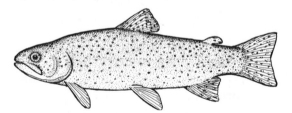

Figure 125.

Bluish on back, silvery on sides and belly. More or less profusely covered with small spots. Dorsal and caudal fins profusely spotted. Length ordinarily 10 to 15 inches. Weight up to six pounds, but some known to reach about 30 pounds. Many sub-species in Rocky Mountain and Pacific coast states ranging into southern Alaska; widely introduced in many areas.

Two closely related species; the Gila trout, *Salmo gilae* Miller, is found in a few remote headwater tributaries of the Gila River in New Mexico and the Arizona trout, *Salmo apache* Miller, is found in the Gila and Little Colorado drainages in Arizona. Both species have a yellowish "cutthroat" mark instead of the red or pink streak on the underside of the mandible. A third species, the Mexican Golden trout, *Salmo chrysogaster* Needham and Gard, is found in the Rio Verde in southwestern Chihuahua, Mexico.

9b **Hyoid teeth absent; lateral line scales usually less than 150; no red or pink streak on underside of mandible; dorsal rays 10-13, usually 11-12 . . . . 10**

10a **Anal fin with 9 rays; adults with small "X"-shaped spots on sides. Fig. 126 . . . . ATLANTIC SALMON,** *Salmo salar* **Linnaeus**

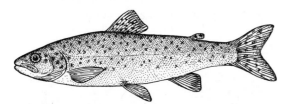

Figure 126.

Bluish brown backs, sides, and belly silvery. Weight up to 10 pounds, but weights up to 100 pounds have been reported in Scotland. Anadromous, Atlantic Ocean, Delaware to Greenland and northern Europe.

Smaller landlocked varieties are found in some eastern lakes.

10b **Anal fin with 10-13 rays; sides usually with round spots except in some old brown trout 11**

11a **Caudal fin without spots or with only a few restricted to dorsal portion. Fig. 127 . . . . BROWN TROUT,** *Salmo trutta* **Linnaeus**

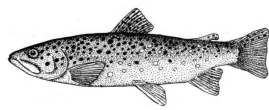

Figure 127.

Dark above, silvery below, back and sides more or less with numerous large spots. Length up to 2 feet. Widely introduced from Europe.

11b **Caudal fin profusely covered with spots 12**

12a **Dorsal, anal, and pelvic fins with some spots, but strongly emarginated with white offset by a dark bar; sides more or less brilliantly colored with yellow and orange; parr marks (vertical bars on sides of juveniles) retained by adults. Fig. 128 . . . . GOLDEN TROUT,** *Salmo aguabonita* **Jordan**

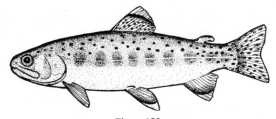

Figure 128.

Length usually 8 to 12 inches, although some may reach a length of 20 inches. Usually confined to streams of high altitudes, originally in the high Sierras of California, but introduced in several other areas. A complex group apparently related to the rainbow trout and containing several forms sometimes regarded as different species in Golden Trout Creek, Little Kern River and the South Fork.

12b Dorsal and anal fins speckled, but not emarginated; side speckled and marked longitudinally with a more or less pinkish streak; no parr marks on adults. Fig. 129 . . . . RAINBOW TROUT, *Salmo gairdneri* Richardson

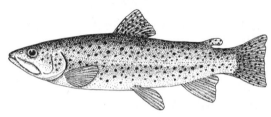

Figure 129.

Bluish on back; sides and belly profusely covered with small spots. Reaches weights of 10 to 15 pounds, but some are known to be much larger. Originally in waters on west slope of the Rockies from Baja California into Alaska, but now widely introduced elsewhere including Hawaii. Many sub-species in western waters, including some anadromous forms.

13a Caudal fin deeply forked; body with light spots but with no red spots. 14

13b Caudal fin not deeply forked; body with light spots and some red spots; lower fins strongly emarginated with whitish or cream color 15

14a Body including fins and head profusely covered with irregular light spots. Fig. 130. . . . . LAKE TROUT, *Salvelinus namaycush* (Walbaum)

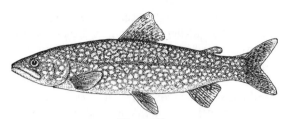

Figure 130.

Color varies from light gray to almost black with profuse light spots. Reaches a length of over 3 feet and a weight of over 80 pounds. The Great Lakes and in colder lakes of the St. Lawrence, Hudson River, and the Great Lake drainages northwestward to headwaters of the Columbia and Fraser Rivers and into Alaska. Widely introduced in western lakes.

CISCOWET, *Salvelinus siscowet* (Agassiz). Fig. 131. Sometimes regarded as a form of the lake trout which it resembles in color and markings but differs largely in its very deep body and thick layer of fat lining the body cavity. Reaches a weight of over 40 pounds. Restricted to the deep waters of Lake Superior where it occurs with the lake trout but does not apparently intergrade.

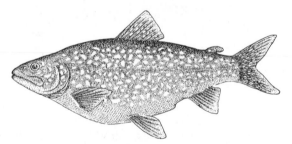

Figure 131.

14b Body profusely covered with light spots which are usually restricted to the sides and are absent from the head and fins. Fig. 132. . . . . ARCTIC CHARR, *Salvelinus alpinus* (Linnaeus)

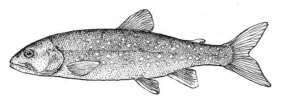

Figure 132.

Body grayish with light spots which may be absent in some landlocked forms. Anadromous, northern Atlantic and northern Arctic drainage of Canada, Gulf and Arctic drainage of Alaska and elsewhere in the Arctic drainage. The blueback trout of Rangely Lakes in Maine once regarded as a different species is currently considered as a landlocked form of the Arctic charr.

15a **Back with mottled or "wormy" streaks on dark background; dorsal and caudal fins mottled. Fig. 133** . . . . **EASTERN BROOK TROUT,**
Salvelinus fontinalis (Mitchill)

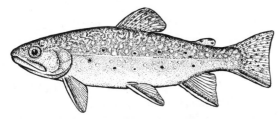

Figure 133.

Back dark olive with somewhat lighter sides and belly (reddish in males). Sides with black and some red spots almost as large as pupil of eye. Dorsal and caudal fins rather mottled. Originally in certain waters from Minnesota eastward, but now distributed elsewhere. Several varieties, even anadromous forms are known.

15b **Back with spots on dark background; dorsal and caudal fins not mottled.**

Fig. 134 . . . . **DOLLY VARDEN,**
Salvelinus malma (Walbaum)

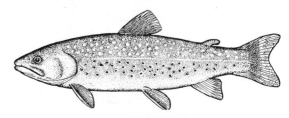

Figure 134.

Rather olivaceous to dark silvery with red spots about the size of eye on sides. Light spots on back. Length up to 20 inches (12 pounds). Streams on west slope of Rocky Mountains from northern California to Alaska and Siberia.

16a **Two flaps on the septum dividing the nostril (Fig. 135); gill rakers of first arch more than 23** 17

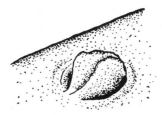

Figure 135.

16b **Single flap on the septum dividing the nostril (Fig. 136); gill rakers of first arch less than 20** 30

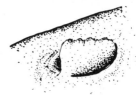

Figure 136.

Key to the Families   49

17a Premaxillary extending downward and backward forming a rounded or blunt snout. Fig. 137. . . . . **18**

Figure 137.

17b Premaxillary extending upward and forward, forming a rather acute or pointed snout. Fig. 138 . . . . **20**

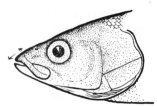

Figure 138.

18a Length of maxillary less than twice its width; gill rakers on first arch 18-25; gill rakers short. . . . . **BROAD WHITEFISH,** *Coregonus nasus* (Pallas)

Olivaceous to brownish above, sides and belly silvery. Scales large, 84-102 in lateral line. Similar to the common whitefish but interorbital width divided by the length of the longest gill raker is more than 0.2 (0.22-0.44). Length up to 18 inches. Fresh and brackish waters of Arctic drainage of northwestern North America, Coppermine, Mackenzie and Yukon Rivers.

18b Length of maxillary more than twice its width; gill rakers on first arch 20-33, usually more than 24; gill rakers long. **19**

19a Mouth inferior, snout overhanging mouth; scales in lateral line usually less than 90 (70-94). Fig. 139 . . . . **COMMON WHITEFISH,** *Coregonus clupeaformis* (Mitchill)

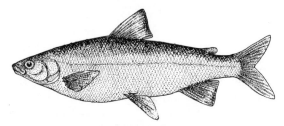

Figure 139.

Olivaceous above, white to silvery below. Interorbital width divided by length of the longest gill raker less than 0.2 (0.13-0.19). Gill rakers 19-33. Length usually 12 to 20 inches, but may exceed 30 inches. Throughout the Great Lakes, various forms or sub-species in many large lakes from New England to Minnesota and northward.

19b Mouth terminal, jaws equal in length; snout not overhanging (snout slightly overhanging in adult males); scales in lateral line 90 or more (90-100) . . . . **ATLANTIC WHITEFISH,** *Coregonus canadensis* Scott

Similar in coloration to the common whitefish. Gill rakers 23-27. Length up to 15 inches. Restricted to the Tusket River system and several lakes in Nova Scotia.

20a Gill rakers on first arch usually less than 30; shape of body pike-like . . . . **INCONNU,** *Stenodus leucichthys* (Güldenstadt)

A slender silvery fish with long pike-like jaws; maxillary extends to middle of eye or beyond (Fig. 140). Length up to 5 feet. Anadromous with migratory landlocked forms. Arctic drainage from Anderson River, Canada, through Alaska to Siberia.

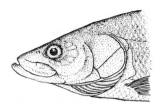

Figure 140.

20b Gill rakers on first arch usually more than 30; shape of body not pike-like **21**

21a Mandible included (Fig. 141) **22**

Figure 141.

21b Mandible terminal (Fig. 142) or prognathus (Fig. 143). **23**

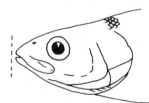

Figure 142.

Figure 143.

22a Lower jaw with considerable black pigment toward the tip; pectoral and pelvic fins relatively short; maxillary extending posteriorly to below the anterior half of eye .... **SHORTNOSE CISCOE,** *Coregonus reighardi* **(Koelz)**

Silvery white below and bluish above. Gill rakers on the first arch 32-42; scales in lateral line 65-83. Length to 15 inches. Lakes Superior, Michigan, Ontario, and Nipigon.

22b Lower jaw without much pigment; pectoral and pelvic fins long; maxillary extending posteriorly to the middle of the eye or beyond. Fig. 144 .... **SHORTJAW CISCOE,** *Coregonus zenithicus* **(Jordan and Evermann)**

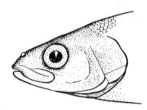

Figure 144.

Silvery with dark back. Gill rakers on first arch 32-46; scales in the lateral line 58-90. Length up to 12 inches. Found in deep waters of Lakes Superior, Michigan, Huron, and Nipigon. Also in Lakes Winnipeg and Athabasca, Canada.

23a Diameter of eye equal to or greater than length of snout. **24**

23b Diameter of eye less than length of snout **25**

24a Gill rakers on the first arch less than 41; diameter of eye equal to snout length. .... **BERING CISCOE,** *Coregonus laurette* **Bean**

Silvery ciscoe with ventral fins immaculate. Gill rakers on first arch 33-40; body depth 20 percent of standard length. Scales in the lateral line 76-95. Length up to 12 inches. Known from coastal waters of Alaska from Cook Inlet to near the Colville River.

24b **Gill rakers on the first arch more than 41; diameter of eye greater than snout length.**
. . . . **LEAST CISCOE,**
Coregonus sardinella **Valenciennes**

Silvery in color, anadromous form with dark spots, spots lacking in non-anadromous forms in fresh water. Gill rakers on first arch 42-53; body depth slightly greater than 20 percent of standard length. Scales in the lateral line 78-98. Length up to 10 inches. Alaska and the Northwest Territories, also in northern Europe and Asia.

25a **Mandible with knob, hook or tubercle at tip** **26**

25b **Mandible without a knob, hook or tubercle at tip.** **27**

26a **Gill rakers on the first gill arch usually less than 38; pectoral and pelvic fins long. Fig. 145.** . . . .
. . . . **KIYI,** *Coregonus kiyi*
(Koelz)

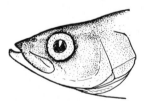

Figure 145.

Silvery with dusky back. Gill rakers on first arch 34-37; thin mandible with prominent symphseal knob. Scales in lateral line 71-91. Small

fish, 6 to 8 inches in length. Deep waters of Lakes Superior, Michigan, and Ontario.

26b **Gill rakers on the first arch usually more than 38 (37-50); pectoral and pelvic fins short. Fig. 146.** . . . .
. . . . **BLOATER,** *Coregonus hoyi* **(Gill)**

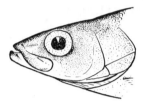

Figure 146.

Silvery with irridescent bluish back. Mandible with tubercle on tip. Scales in lateral line 63-84. Length 6 to 8 inches. Deep waters of Lakes Superior, Michigan, Huron, Ontario, and Nipigon.

27a **Body deepest in front of center** **28**

27b **Body deepest at center** **29**

28a **Gill rakers on first arch 36 or fewer** . . . . .
. . . . **DEEPWATER CISCOE,**
Coregonus johannae **(Wagner)**

Silvery fishes with dusky backs. Gill rakers 25-36; scales in lateral line 68-83. Deep waters of Lakes Michigan, Huron, and Erie, Parsons and Todd state that the species may be extinct.

28b **Gill rakers on first arch more than 36. Fig. 147.** . . . .
. . . . **BLACKFIN CISCOE,**
Coregonus nigripinnis **(Gill)**

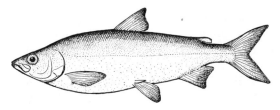

Figure 147.

Dark bluish above and silvery below. Pectoral, pelvic and anal fins darkly pigmented. Gill rakers on first arch 41-54; scales in lateral line 78-89. Reached a length of over 20 inches. Deep waters of Lakes Superior, Michigan, Huron, Ontario, and Nipigon. Like the deepwater ciscoe the blackfin ciscoe has not been reported from the lakes in recent years and may be extinct.

29a Gill rakers on the first arch usually 44 to 64 . . . .
. . . . CISCOE, *Coregonus artedi* Lesueur
Silvery below shading into a more or less dark bluish on back. Length usually 12 to 15 inches, but may exceed 24 inches. Gill rakers 38-64, usually more than 44; scales in the lateral line 63-94. Many forms or sub-species have been described, the ciscoe or lake herring (Fig. 148) in

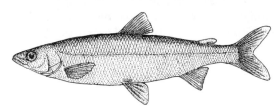

Figure 148.

the Great Lakes and the tullibee (Fig. 149) is known from many large and deep lakes of the northern United States and Canada. Those found in the Great Lakes tend to be more slender.

A similar and perhaps closely related species, the Arctic ciscoe, *Coregonus autum-*

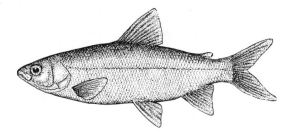

Figure 149.

nalis (Pallas), occurs in the lower reaches of Arctic rivers in Alaska and the Northwest Territories. Another similar species *C. nipigon* is found in Lake Nipigon. The latter species is thought by some workers to be a sub-species or form of *C. artedi*.

29b Gill rakers on the first arch usually less than 44 . . . .
. . . . LONGJAW CISCOE, *Coregonus alpenae* (Koelz)
Silvery with a greenish or bluish back, sides and ventral surface silvery. Little or no pigment on the fins. Gill rakers on first arch 30-46; scales in the lateral line 68-83. Length up to 10 inches. A deepwater ciscoe formerly found in Lakes Michigan, Huron, and Erie but it has not been collected for two decades and is probably extinct.

30a Scales in lateral line less than 75, usually 55-60; size small, length usually less than 8 inches. Fig. 150. . . . .
. . . . PYGMY WHITEFISH, *Prosopium coulteri* (Eigenmann and Eigenmann)

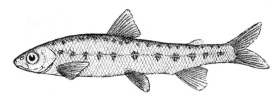

Figure 150.

Dull silvery with a dusky back. The headwaters of the Columbia River into Alaska. Also found in the deep waters of Lake Superior.

BEAR LAKE WHITEFISH, *Prosopium abyssicola* (Snyder). Very similar to pygmy whitefish but differs in larger number of scales in lateral line, 67-78 instead of 59-65. Bear Lake drainage, Idaho and Utah.

30b   Scales in lateral line more than 75; fish larger; length usually more than 8 inches **31**

31a   Tip of snout below level of eye; profile of head rounded. Fig. 151 . . . . **MOUNTAIN WHITEFISH,** *Prosopium williamsoni* **(Girard)**

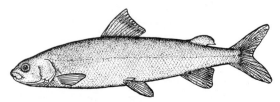

Figure 151.

Bluish above and silvery white below; all fins tipped with black. Length up to 15 inches. Lakes and streams of western North America. Fraser and Columbia River systems in the Yukon and British Columbia north to the Laird River. The Truckee River and Lahontan basin in Nevada. Idaho, Wyoming, Montana, and Saskatchewan and Milk Rivers in Alberta.

BONNEVILLE WHITEFISH, Prosopium spilonotus (Snyder) is very similar and may be a sub-species. It is restricted to Bear Lake Basin, Idaho and Utah.

BONNEVILLE CISCOE, *Prosopium gemmiferum* (Snyder). Appearance similar to the lake herring; length about 7 inches. Restricted to Bear Lake drainage, Idaho and Utah.

31b   Tip of snout below level of eye; profile of head not rounded. Fig. 152 . . . . **ROUND WHITEFISH,** *Prosopium cylindraceum* **Pallas**

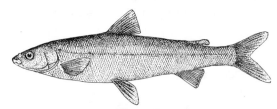

Figure 152.

Dark blue above and light silvery below. Length up to about 15 inches. Two disjunct populations exist. The eastern population from northern New Brunswick to Labrador, west to Ungava; Maine south to New York and west through the Great Lakes to Lakes Superior and Nipigon. A western population from northern Manitoba northwestward through the Northwest Territories, Yukon, northern British Columbia to Alaska.

SMELT FAMILY
Osmeridae

The American species of the smelt family, Osmeridae, are all found in the sea, but a number of species are anadromous and enter rivers in large spawning runs in the spring. The rainbow smelt, *Osmerus mordax* (Mitchill), has become landlocked in some eastern lakes, and was accidentally introduced years ago into the Great Lakes where it now flourishes.

These are all small predaceous fishes seldom reaching a length of more than ten or twelve inches, but some species are highly prized for food.

In general they have slender, silvery bodies and are characterized by adipose fin, large scales and strong jaws with well-developed teeth. They do not have the axillary process found at the base of the pelvic fin which is present in some of the related families possessing the adipose fin.

They are predaceous, feeding on all sorts of small aquatic animals, including small fishes.

1a   **Teeth on vomer few and rather large (canine-like)** 2

1b   **Teeth on vomer numerous and small (not canine-like)** 3

2a   **Vomerine teeth moderate size, not fang-like; front of dorsal fin, definitely behind front of pelvic fins. Fig. 153** .... **EULACHON,** *Thaleichthys pacificus* **(Richardson)**

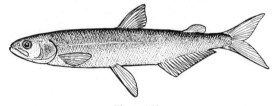

Figure 153.

Dark above and silvery white below. Teeth shed during spawning season. Very oily and reputed to have been used as candles. Length up to 12 inches. Anadromous, entering streams on the Pacific coast from California northward into Alaska.

2b   **Vomerine teeth consisting of 1 to 3 fang-like teeth on each side of the tip of the vomer; front of dorsal fin over or above front of pelvic fins. Fig. 154** .... **RAINBOW SMELT,** *Osmerus mordax* **(Mitchill)**

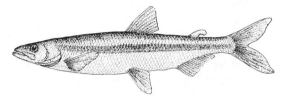

Figure 154.

The rainbow smelt is represented by two sub-species, the sub-species *mordax* is the common smelt of the eastern states and Canada and the sub-species *dentax* occurs in the Pacific from Vancouver Island north along the Alaskan coast to the Mackenzie River delta region in the Canadian Arctic. Rather colorful with a pale olive back, sides purple above lateral line shading below to blue, violet and gold to white on the belly. Length up to 12 inches, landlocked populations tend to be smaller 6-8 inches. Anadromous, entering streams from New Jersey to Labrador and from British Columbia into Alaska and the Canadian Arctic to the Mackenzies basin, the Aleutian Islands and Siberia. Landlocked in some waters of New York, New England, New Brunswick, Nova Scotia, Newfoundland, and in the Great Lakes. Recently introduced into several lakes in the Hudson Bay drainage and St. Croix River drainage in Minnesota.

3a   **Mouth large, 1.8 to 2.2 times in head length; maxillary extending to or beyond posterior margin of pupil; teeth large, but in one row on vomer and palatine** .... **LONGFIN SMELT,** *Spirinchus thaleichthys* **(Ayres)**

Silvery with rather dusky back. Length up to 12 inches. Marine but may enter fresh water from California to British Columbia.

NIGHT SMELT, *Spirinchus starski* (Fisk) is marine but has been reported to enter mouths of rivers, Monterey, California to Washington.

3b Mouth small, 2.2 to 2.5 times in head length; maxillary extending not behind center of pupil of eye; teeth small and in 2 rows on vomer and palatine. . . . . POND SMELT, *Hypomesus olidus* (Pallas)

Silvery with dark back. Length up to 12 inches. Anadromous, entering the streams on the Pacific coast from Alaska to Japan.

SURF SMELT, *Hypomesus pretiosus*

(Girard). Differs in having a longer snout, more than 65 lateral line scales and the front of the dorsal fin before the front of the pelvic fins. Marine, found from Monterey, California northward into Alaska, sometimes entering fresh water.

DELTA SMELT, *Hypomesus transpacificus* McAllister. Fig. 155. A brackish and freshwater species formerly confused with the pond smelt. Lower Sacramento and San Joaquin Rivers, California.

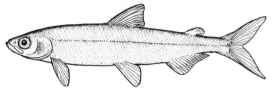

Figure 155.

PIKE FAMILY
Esocidae

This family includes the muskellunge and the several species commonly known as various kinds of pike and pickerel, all belonging to the genus *Esox*. Very few groups of fishes have received as many common names as the several species comprising this family.

The members of this family are characterized by long cylindrical bodies with prominent jaws shaped like a duck's bill and armed with numerous fang-like teeth. They possess a soft dorsal fin which is located far back on the body.

Pickerel and pike vary in color from an olive brown to a pale silver with light undersides. They are marked with light or dark bars or spots depending on the species.

The several species of pike and pickerel are widely distributed through northern North America, Asia, and Europe. They are highly predaceous, feeding on fishes and any other liv-

ing animals small enough to seize. They spawn in the spring, scattering their eggs at random in shallow water where the eggs are fertilized and left to develop without any parental care. Pike grow very rapidly and some species reach a large size. The larger species are popular game fishes.

1a Opercle with scales on upper half only (Fig. 156). 2

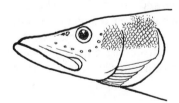

Figure 156.

1b Opercle entirely scaled (Fig. 157) 3

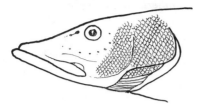

Figure 157.

2a Pores on each side of ventral surface of lower jaw, six or more (Fig. 158); body with dark vertical bars or spots or with no marks on light background; lower half of cheek usually wholly or partially without scales. Fig. 159 . . . . **MUSKELLUNGE**, *Esox masquinongy* Mitchill

Figure 158.

Figure 159.

Reaches a length of about 60 inches and weights up to 75 pounds are known. Three geographical races exist as follows; St. Lawrence and lower Great Lakes drainage; upper Ohio Valley; northwestern Wisconsin, northern Minnesota and adjacent Ontario.

2b Pores on each side of ventral surface of lower jaw, 5 or less (Fig. 160); body with small light spots on dark background (vertical bars in juveniles); cheek always entirely scaled (Fig. 161); variant known as silver pike has lost all body spots. Fig. 162 . . . . **NORTHERN PIKE**, *Esox lucius* Linnaeus

Figure 160.

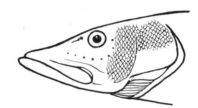

Figure 161.

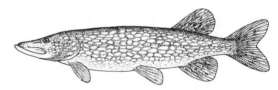

Figure 162.

Reaches a length of over 40 inches and a weight of over 30 pounds. Eastern U.S. north of the Ohio River and northwestward into Alaska and in northern Siberia and Europe.

3a Sides and back marked with dark network; scales of lateral line 125; branchiostegal rays 14-16. Fig. 163 . . . .

.... **CHAIN PICKEREL,**
Esox niger **Lesueur**

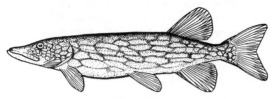

Figure 163.

This is the common pickerel of the New England States. It reaches a length of over 24 inches and a weight of 10 pounds. East of the Appalachians from the St. Lawrence southward and along the Gulf coast to Texas.

3b Sides and back marked with dark wavy or wormy vertical streaks; lateral line scales 105; branchiostegal rays 11-13. Fig. 164

.... **REDFIN PICKEREL,**
Esox americanus **Gmelin**

Figure 164.

Very small, seldom reaching a length of more than 12 inches. Atlantic coastal plain, Maine to Florida.

The grass pickerel, *Esox americanus vermiculatus* Lesueur is similar in appearance and size to the redfin pickerel but the head is longer. Intergrades occur from Iowa southeastward through the Ohio Valley and south to the Gulf states from Texas to Florida.

MUDMINNOW FAMILY
Umbridae

This family contains three genera with four American species. Mudminnows are soft-rayed and have the dorsal fin rather far back. They are more or less reddish brown and may be somewhat mottled. The lateral line is absent. Their upper jaw is non-protractile, lacking a groove separating it from the snout.

Mudminnows commonly inhabit swamps, muddy streams, and sloughs where they often bury themselves in the mud. They are hardy and have a remarkable resistance to drought and winter conditions. Mudminnows spawn in the early spring. They are reputed to be omnivorous, but feed heavily on small aquatic insects and crustaceans.

1a Front of anal fin almost immediately under front of dorsal fin; scale rows more than 50 (70). Fig. 165. ....
.... **ALASKA BLACKIFISH,**
Dallia pectoralis **Bean**

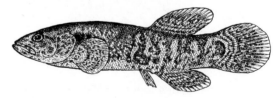

Figure 165.

Brownish mottled with white. Arctic and central Alaska and St. Matthew and St. Lawrence Islands. Length 8 inches.

1b Front of anal fin behind front of dorsal fin 2

2a Scale rows more than 50; anal fin rays 10-11. Fig. 166. . . . . WESTERN MUDMINNOW, *Novumbra hubbsi* Schultz

Figure 166.

Very similar to the central mudminnow in appearance, but differs in having a supermaxillary. Length 4 inches. Chehalis River, Deschutes River at Olympia, Cook Creek (Quinault River) drainages, Washington.

2b Scale rows less than 50; anal fin rays 7-8. 3

3a Body with longitudinal streaks. Fig. 167 . . . . EASTERN MUDMINNOW, *Umbra pygmaea* (DeKay)

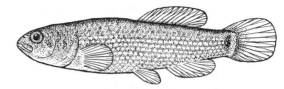

Figure 167.

Length 3 inches. Swamps and sluggish streams of Atlantic coastal plain, Long Island to Florida.

3b Body without longitudinal streaks, but with more or less faint crossbars. Fig. 168. . . . . CENTRAL MUDMINNOW, *Umbra limi* (Kirtland)

Figure 168.

Length usually about 2 inches, but sometimes exceeding 6 inches. Swamps and sluggish streams of upper Mississippi valley and Great Lakes region and into Manitoba.

CHARACIN FAMILY
Characidae

The Characins form a large family of about 300 species found mostly in South America and Africa. They possess an adipose fin. They have a wide range of form and habits.

Some species extend up into Mexico, and, at least, one species ranges to the Rio Grande River. This is the Mexican tetra, *Astyanax fasciatus mexicanus* (Filippi) (Fig. 169) which is

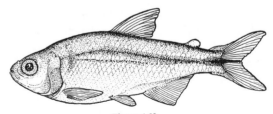

Figure 169.

found in southern Texas, and also in the lower Colorado River drainage of Arizona and New Mexico. It has been introduced elsewhere in the southwestern United States. It reaches a length of about 4 inches.

MINNOW FAMILY
Cyprinidae

The minnow family is the largest family of freshwater fishes in North America and contains more species and more individuals than any other family. Only fishes belonging to this family are true minnows, although the name is often incorrectly applied to some small fishes in other families. Most of our minnows are small, but that is not true of all as several species reach a length of three to five feet.

Minnows are closely related to the sucker family, but differ in several respects. In general, they lack the sucker-like mouth, although several species of minnows have mouths that closely resemble those of suckers. Both suckers and minnows lack teeth in the mouth, but both have well-developed teeth on the last or modified fifth gill arch, the pharyngeal arch. However, the pharyngeal teeth of the minnows are not as numerous as those of the suckers and may have more than one row of teeth, usually one or two rows in our native minnows (Fig. 170) or three rows in the introduced carp (Fig. 171).

Minnows lack spines except for hardened rays in several introduced forms, carp and goldfish, and in several peculiar species in the deserts of the southwest. Most minnows have less than 10 dorsal fin rays, usually 8 or 9 rays, although there are several species including the carp and goldfishes that have more than 10.

Minnows are most important fishes. Their great abundance and small size make them valuable as food for other fishes. Various

Figure 170.

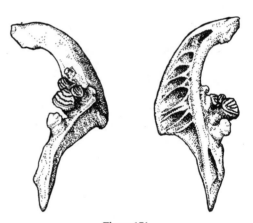

Figure 171.

minnows utilize the minute animal and plant life of our waters, and in turn furnish food for the game fishes. Many species have long been utilized as bait, and at present the bait industry has become a major industry in many parts of the United States. A few of the larger minnows are used as food. The carp, originally introduced as a game and food fish, have supported a large fishery industry ever since their successful establishment in North America. Minnows have a complete series of multiple ribs from the posterior margin of the head to the tip of the tail which makes them excessively bony. The ribs together with the coarse flesh and muddy flavor of many species does not make them very popular for food. Our minnows have never achieved the status of sport fish that their relatives in the Old World have attained.

Minnows feed on almost every type of food found in our rivers, streams, and lakes. Some are mud feeders, others are vegetarians, and many are predators upon the various forms of minute animal life. Many species are omnivores and feed on plants, animals, and detritus. The larger minnows may be important predators on other smaller minnows and the young of other families of fish.

Most species of minnows spawn either in the spring or early summer, but the spawning habits of the various species differ considerably. Some, such as the carp and the western squawfish, crowd into shallow waters and deposit their eggs with great splashing. Other species have quite elaborate behavior patterns, the male may build a nest of small stones in which the female deposits her eggs and they are fertilized by the male. The male usually guards the nest from intrusions by other males of the same species. Once spawning is completed the males leave the nest and the eggs hatch unattended. Many closely related species may use the same section of the stream for spawning and occasionally the sperm of one species may fertilize the eggs of another species and hybrids may result. In fact hybrids are quite common between various species of minnows and these hybrids may be very difficult to identify, as well as, very confusing. Other species of minnows deposit their eggs on the underside of a log or stone and the males guard and clean the eggs until they hatch.

Minnows are difficult to key to species. Most keys use the type and number of teeth on the pharyngeal arch as one of the main diagnostic characters. In the following keys we have attempted to use characters other than the type and numbers of pharyngeal teeth and this has resulted in an artificial separation of the various genera and species. We have listed the pharyngeal tooth formula in the short species description following the species name. One of the more important characters useful in the identification of the species of cyprinids is the pharyngeal teeth, their numbers, shapes, and arrangement. The teeth are arranged in up to three rows on each posterior branchial or pharyngeal arch. As mentioned previously native North American minnows have teeth arranged in not more than two rows, the introduced carp has teeth in three rows. In our native minnows the outer row may have 0, 1, or 2 teeth and the inner or main row 4 or 5 teeth. The tooth formula for the blacknose shiner, *Notropis heterolepis,* is written 0,4-4,0, which means that the pharyngeal teeth are in a single row or a main row of four teeth on each arch. The blackchin shiner, *Notropis heterodon,* has a tooth formula of 1,4-4,1, indicating the presence of one tooth in the outer left row, four teeth in the inner left and right row and one tooth in the outer right row. Usually the tooth arrangement on the pharyngeal arches is symmetrical, but there are exceptions. As an example in the creek chub, *Semotilus atromaculatus,* the teeth are in two rows, either 2,4-5,2 or 2,5-4,2.

Pharyngeal teeth like other morphological characters are variable. In some species, e.g., the spottail shiner, *Notropis hudsonius,* the typical formula is 2,4-4,2, but several other arrangements are found, 2,4-4,1; 1,4-4,2; 0,4-4,1; etc. Such variability is unusual but in many instances two or more tooth formulae may be found and in such instances the least common will be placed in parenthesis, e.g., (1)2,4-4,2(1) or (0)1,4-4,1(0). The pharyngeal teeth are deciduous and are lost and replaced at irregular intervals. If teeth are missing their positions are marked by empty sockets that can be easily seen with the aid of a dissecting microscope or a 10X hand lens.

In order to see the pharyngeal teeth the paired pharyngeal bones, each of which is formed by the fusion of two dorsal bones with a single ventral bone, must be removed. The dissection is not difficult but must be carried out with care or teeth may be accidentally broken off or the arch itself may be fractured. Turn back the operculae and expose the last pharyngeal arch. The arch lies closely pressed to and inside the shoulder girdle supporting the pectoral fins. The arch must be separated from the girdle by cutting each side dorsally and ventrally severing the muscles and connective tissue, then the arch can be removed by cutting each end. A narrow thin-bladed scalpel is useful for this purpose. A set of small needle-nose forceps, straight and curved, are also useful in removing the separate arches. Once the arches are out, the tissues must be dissected away from the teeth if the teeth are not visible. Fine needles and forceps should be used, being careful not to remove the very fragile teeth. Fine sewing needles inserted into wooden handles are easily constructed and useful for teasing away the tissues.

The pharyngeal arches are most easily removed from fresh unfixed specimens. After preservation the heavy muscles attached to the arches are less easily removed. When the tooth counts have been completed the arches from the specimen should be replaced so they will be available to other investigators. If the collection consists of a single specimen the arches can be placed in a small vial in the container with the specimen. The entire procedure, while initially difficult and at times quite frustrating, can be mastered with a little practice. In fact it is wise to carry out practice dissections on larger more common minnows before attempting to remove the arches from smaller specimens.

In addition to normal variability within a species there are considerable morphological differences in the teeth of different species of minnows, ranging from the heavy molariform grinding teeth of the carp to the fragile strongly hooked teeth found in species of the genus *Notropis*. Tooth morphology gives important clues to the food habits of the species. Molariform teeth are characteristic to herbivores and mollusc feeders and are used to grind up plant material or to crush snails or small bivalves. Hooked teeth, with or without serrations, are characteristic of predators that feed on insects and other benthic animals. The sharp teeth pierce the tough exoskeleton of arthropods or may serve to tear up the prey. The effectiveness of pharyngeal teeth can be readily appreciated by placing the index finger carefully in the pharynx of a larger minnow such as a creek chub or carp.

1a   **More than 12 dorsal rays** 2

1b   **Less than 12 dorsal rays** 3

2a   **With 2 barbels on each side of upper jaw; anterior spine in dorsal and anal fin** . . . . *Cyprinus*
(EUROPEAN) CARP, *Cyprinus carpio* Linneaus. Fig. 172. Reddish brown on back to silver below. Length up to 30 inches. Widely introduced in the U.S. from Europe, originating in Asia. Common in Hawaii but absent in Alaska.

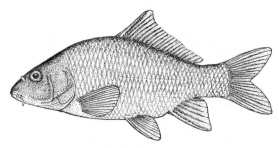

Figure 172.

2b Without barbels on upper jaw; anterior spine in dorsal and anal fin . . . . *Carassius* GOLDFISH, *Carassius auratus* (Linneaus). Fig. 173. Color varies from reddish gold to brown. Length up to 12 inches. Introduced and liberated in many parts of U.S., including Hawaii.

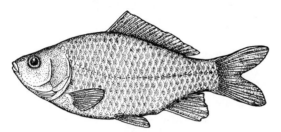

Figure 173.

3a Dorsal fin without spines 4

3b Dorsal fin with double spines 84

4a Lower jaw with distinct inner cartilaginous or horny ridge. (Do not confuse with soft flap in other fishes.) Fig. 174 5

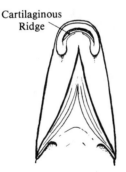

Cartilaginous Ridge

Figure 174.

4b Lower jaw without a distinct inner cartilaginous ridge 9

5a Cartilaginous or horny ridge confined to lower jaw 6

5b Cartilaginous or horny ridge in both upper and lower jaw 8

6a Intestine wrapped many times around swim bladder; sum of lateral line scales and circumferential scales less than 105 7

6b Intestine rarely wrapped around intestine; sum of lateral line scales and circumferential scales more than 105 . . . . MEXICAN STONEROLLER, *Campostoma ornatum* Girard

Similar to the large-scale and common stoneroller, except the intestine encircles the air bladder in only a small percentage (20%) of specimens. The scales are smaller, 58-77 in the lateral line and 47-60 circumferential scales; snout somewhat acute. Big Bend Region of Texas, Rucker Canyon, Arizona and widespread in northern Mexico.

7a Sum of lateral line and circumference scales more than 83 (Figs. 175, 176) . . . . COMMON STONEROLLER, *Campostoma anomalum* (Rafinesque)

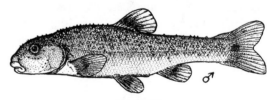

Figure 175.

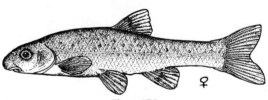

Figure 176.

7b Sum of lateral line and circumference scales less than 83 .... LARGE-SCALE STONE-ROLLER, *Campostoma oligolepis* **Hubbs and Greene**

Similar to the common stoneroller, tanish to slate gray, little mottled. Breeding males anal fin without dark crossbar; base of dorsal fin usually whitish; about 45 scales in the lateral line. Southeastern Minnesota, eastern Iowa, Wisconsin, northern Illinois, Missouri and Ozark Upland of Arkansas.

8a Dorsal rays about 10; ridges inside of jaws covered by heavy and not easily removed horny sheaths. (Fig. 177). .... *Acrocheilus*

Figure 177.

CHISELMOUTH, *Acrocheilus alutaceus* Agassiz and Pickering. Fig. 178. Dark, belly somewhat lighter; caudal fin very long. Teeth 4-5. Length 12 inches. Columbia River drainage of Washington, Oregon, Nevada, and British Columbia.

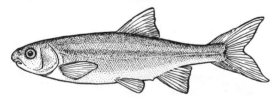

Figure 178.

8b Dorsal rays usually 8 (7-8); ridges inside of jaws covered by easily removed horny sheaths. .... *Eremichthys*
DESERT DACE, *Eremichthys acros* Hubbs

and Miller. Fig. 179. Warm Springs, Soldier Creek, Nevada.

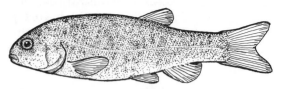

Figure 179.

9a Lateral line scales more than 100; a barbel present at end of maxillary .... *Tinca*
TENCH, *Tinca tinca* (Linneaus). Fig. 180. A fine scaled minnow introduced from Europe; tooth formula 0,(4)5-5,0. Western parts of the U.S., Connecticut, Delaware, and Maryland.

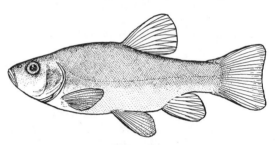

Figure 180.

9b Lateral line scales less than 100; if more than 100, the barbel is absent 10

10a Premaxillaries (protractile) separated from snout by a complete groove (caution: an apparently complete groove may not be complete but may be connected by a hidden bridge of skin) (Fig. 181) 16

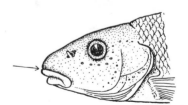

Figure 181.

10b **Premaxillaries (not protractile) not separated completely from head, but connected by a bridge of skin or frenum to tip of snout; this bridge may be a slight connection hidden in the groove in several western species. (Fig. 182)** 11

Figure 182.

11a **Lower lip not modified by side lobes but continuous with central part** 12

11b **Lower lip modified by fleshy side lobes separate from the central part. (See Fig. 192.)** 14

12a **Small barbel usually at end of maxillary. (See Fig. 202.)** .... *Rhinichthys*
Most of the various forms described for this genus can be reduced to sub-species of five species. They tend to have the sides mottled. The barbel may fail to develop or may be very minute, also a frenum may sometimes be minute or absent which may render them difficult to key. See couplet 52a.
BLACKNOSE DACE, *Rhinichthys atratulus* (Herman). Fig. 183. More or less

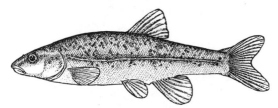

Figure 183.

blackish above, silvery below with lateral band (rosy on sides of male); snout rather short scarcely overhanging upper lip (Fig. 184).

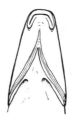

Figure 184.

Frenum sometimes absent. Lateral line scales usually less than 65. Teeth 2,4-4,2. Length 4 inches. Several sub-species in streams from North Dakota to the St. Lawrence drainage and south to Nebraska and North Carolina.

SPECKLED DACE, *Rhinichthys osculus* (Girard). Fig. 185. Similar to blacknose dace but has poorly developed lateral band and more than 55 lateral line scales. Teeth 2,4-4,2. Slight frenum connecting premaxillaries to snout sometimes present. Many sub-species west of the Rockies in coastwide streams of Washington and in Columbia River drainage and south to southern California, Lahontan Basin, and Colorado River drainage.

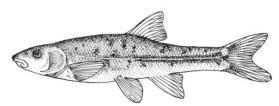

Figure 185.

LEOPARD DACE, *Rhinichthys falcatus* (Eigenmann and Eigenmann). Fig. 186. Resembles other members of this genus but has a falcate dorsal fin and lacks a frenum which may key this species elsewhere. See couplet 52a. Lower Columbia River drainage.

Figure 186.

LONGNOSE DACE, *Rhinichthys cataractae* (Valenciennes). Fig. 187. More or less blackish above, silvery below; back and sides rather

Figure 187.

mottled; snout very elongated (Fig. 188). Spring males reddish on sides. Teeth 2,4-4,2. Length 5 inches. Many sub-species over most of the U.S. and northwestern Canada except Alaska and the southeastern coastal region.

Figure 188.

UMPQUA DACE, *Rhinichthys evermanni* Snyder. Similar to longnose dace but differs partly in having 9 dorsal rays instead of 8. Umpqua River drainage, Oregon.

12b No barbel at end of maxillary. 13

13a Lateral line scales about 75; maxillary extends to front of eye . . . *Mylopharodon*
HARDHEAD, *Mylopharodon conocephalus* (Baird and Girard). Fig. 189. Dark above, pale below. Teeth 2,4-5,2. Large size, length to 3 feet. Sacramento River system, California.

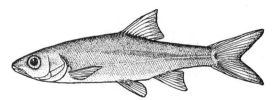

Figure 189.

MOAPA DACE, *Moapa coriacea* Hubbs and Miller. Specimens with a slight frenum or bridge of skin connecting premaxillary to snout will key here. (See Fig. 235, couplet 50a.)
LOACH MINNOW, *Tiaroga cobitis* Girard. If lips are considered to be without side lobes, specimens will key here. (See Fig. 191, couplet 14a.)

13b Lateral line scales about 100; maxillary does not reach front of eye. . . . *Orthodon*
SACRAMENTO BLACKFISH, *Orthodon microlepidotus* (Ayres). Fig. 190. Plain olivaceous above, light below. Teeth 6-6(5). Length 12-16 inches. Sacramento River system, California.

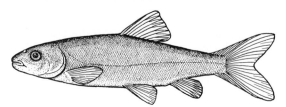

Figure 190.

14a Scales well developed mostly along lateral line; central part of lower lip fleshy . . . . *Tiaroga*
LOACH MINNOW, *Tiaroga cobitis* Girard. Fig. 191. Slender body; olivaceous with small caudal spot; pair of yellowish spots at base of caudal fin. Lower lips are very thick and lateral creases give appearance of side lobes. Teeth 1,4-4,1. Length 2 1/2 inches. Gila River system, Arizona, and New Mexico.

Figure 191.

14b Scales well developed on most of body; central part of lower lip not fleshy 15

15a Central part of lower tip protrudes like a tongue (Fig. 192); no barbel present. Fig. 193. . . . . **CUTLIPS MINNOW,** *Exoglossum maxillingua* (Lesueur)

Figure 192.

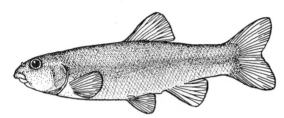

Figure 193.

Dusky olivaceous above, lighter below. Teeth 1,4-4,1. Length 6 inches. St. Lawrence and Lake Ontario south into Virginia.

15b Central part of lip thin, but not protruding (Fig. 194). May or may not have a barbel. . . . . **TONGUETIED MINNOW,** *Exoglossum laurae* (Hubbs)

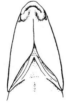

Figure 194.

Similar to the cutlips minnow. Kanawha River drainage of Virginia, West Virginia, and North Carolina and the Allegheny River system.

16a Lower lip with fleshy side lobes; mouth sucker-like (Fig. 195) . . . . . . *Phenacobius*

Figure 195.

SUCKERMOUTH MINNOW, *Phenacobius mirabilis* (Girard). Fig. 196. Pale olivaceous with a lateral band; black spot at base of caudal fin; breast naked; teeth 4-4. Length 4 inches. Colorado and South Dakota to western Ohio, Louisiana, and Texas.

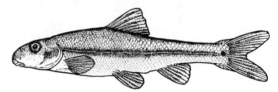

Figure 196.

KANAWHA MINNOW, *Phenacobius teretulus* Cope. Fig. 197. Without a distinct caudal spot and with a scaly breast. Kanawha River drainage in Virginia and West Virginia.

Figure 197.

STARGAZING MINNOW, *Phenacobius uranops* Cope. Fig. 198. Narrow dark lateral band ending in distinct caudal spot. Breast and belly naked. Green, Cumberland and upper Tennessee River drainages.

Figure 198.

RIFFLE MINNOW, *Phenacobius catostomus* Jordan. Fig. 199. Faint dorsal stripe; heavy dark lateral band and indistinct caudal spot. Breast naked, belly partly or entirely scaled. Length 4 inches. Alabama River system.

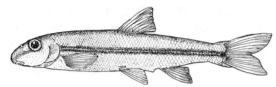

Figure 199.

FATLIPS MINNOW, *Phenacobius crassilabrum* Minckley and Craddock. Fig. 200. Distinct caudal stripe; lateral band diffuse anteriorly and caudal spot indistinct. Mouth broad, pelvic fins reach to anus; belly with small scales. Length 4 inches. Upper Tennessee River drainage.

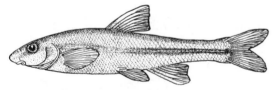

Figure 200.

16b Lower lip without fleshy side lobes; mouth not sucker-like 17

17a Barbel present on side or tip of maxillary; may be large but often minute and occasionally undeveloped in some individuals. 18

17b Barbel absent from maxillary. 45

18a Barbel a short distance in front of end of maxillary and often concealed in groove between maxillary and premaxillary (Fig. 201). 19

Barbel in Front of End of Maxillary
Figure 201.

18b Barbel located at posterior end of maxillary (Fig. 202) 20

Barbel at End of Maxillary
Figure 202.

19a Upper jaw reaches to or behind front of eye; lateral line scales less than 60; dark silvery color . . . . *Semotilus*
CREEK CHUB, *Semotilus atromaculatus* (Mitchill). Fig. 203. Bluish above, light below; adults with black spot at anterior base of dorsal fin. Spring males rosy on sides. Teeth 2,5-4,2. Length up to 10 inches. Montana and Manitoba to eastern Canada and south to the Gulf.

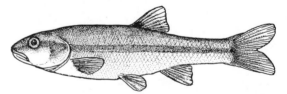

Figure 203.

FALLFISH, *Semotilus corporalis* (Mitchill). Fig. 204. Similar to the creek chub, but lacks black spot in dorsal fin. Eastern Canada and

James Bay drainage to New Brunswick and south on east side of Appalachians to Virginia.

Figure 204.

19b Upper jaw does not reach to eye; color dark and sometimes blotched. Fig. 205 .. . . **PEARL DACE,** *Semotilus margarita* (Cope)

Figure 205.

Dusky mottled above and partly on sides, silvery below. Reddish in spring males. Scales 52-58 in lateral line. Teeth 2,5-4,2. Length 4 inches. Eastern Great Lakes drainage and Vermont southward to Virginia east of the Alleghenies. Sub-species *S. m. nachtriebi* (Cox) with smaller scales, 65-75, northwest Canada and northern states from the Rockies to New York.

20a Lateral line scales more than 70 . . . . . . **21**

20b Lateral line scales less than 70 . . . . **22**

21a Maxillary does not reach eye; teeth in two rows . . . . *Mylocheilus* PEAMOUTH, *Mylocheilus caurinus* (Richardson). Fig. 206. Dark above, silvery on sides with dark lateral band below which is a shorter dark band. Males are reddish on belly and sides. Teeth (1)2,5-5,2(1). Length over 12 inches. Lower Columbia River drainage.

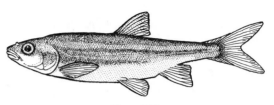

Figure 206.

21b Maxillary reaches eye; teeth in one row . . . . *Agosia* LONGFIN DACE, *Agosia chrysogaster* Girard. Fig. 207. Similar to *Rhinichthys* but with a slight frenum hidden in the groove of premaxillary which is easily overlooked. Dark above, pale below with lateral band; male may have orange sides. Teeth 4-4. Length 4 inches. Lower Colorado River drainage.

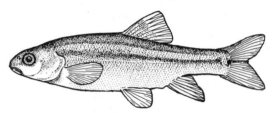

Figure 207.

22a Dorsal lobe of caudal fin much longer than ventral lobe . . . . *Pogonichthys* SACRAMENTO SPLITTAIL, *Pogonichthys macrolepidotus* (Ayres). Fig. 208. Silvery; rudimentary rays of caudal fin very well developed. Teeth 2,(5)4-5,2. Length over 12 inches. Sacramento River system of California.

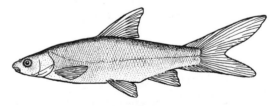

Figure 208.

CLEAR LAKE SPLITTAIL, *Pogonichthys ciscoides* Hopkirk, similar to the Sacramento Splittail, differing in having more gill rakers, 18-23 versus 14-18, and more scales in the lateral line, 60-69 versus 57-64. A ciscolike fish found in Clear Lake, California.

22b Dorsal lobe of caudal fin not longer than ventral lobe 23

23a Snout not bulging forward or extending to any appreciable distance beyond upper lip (Fig. 209); mouth more or less terminal. 24

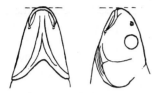

Figure 209.

23b Snout bulging or distinctly extending beyond upper lip (Fig. 210); mouth distinctly sub-terminal 31

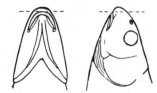

Figure 210.

24a Scales small, lateral line scales more than 50. Fig. 211 . . . .
. . . . **LAKE CHUB,**
Couesius plumbeus (Agassiz)

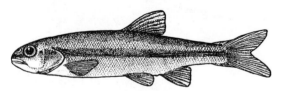

Figure 211.

Dusky color with a rather obscure lateral band. Barbel slightly anterior to end of maxillary. Teeth 2,4-4,2. Length up to 6 inches. Macken-

zie River basin, most of Great Lakes, Hudson River, and Delaware River drainages. Represented by sub-species west of Lake Superior.

24b Scales larger, 50 or less in the lateral line 25

25a Snout about equal to diameter of eye; caudal spot mostly on rays of the caudal fin 26

25b Snout much longer than diameter of eye; caudal spot mostly on peduncle at base of caudal fin 27

26a Body deep; caudal spot faint. Fig. 212 . . .
. . . . **OREGON CHUB,**
Hybopsis crameri Snyder

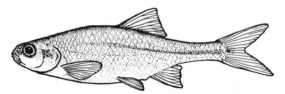

Figure 212.

Somewhat speckled or mottled; barbel minute and sometimes poorly developed. Length about 2 inches. Willamette and Umpqua Rivers, Oregon.

26b Body slender; caudal spot prominent and black. Fig. 213 . . . .
. . . . **REDEYE CHUB,**
Hybopsis harperi (Fowler)

Figure 213.

Silvery below with dark lateral band above which is a light streak, and with a dark back, lateral band extends around snout. Length about 2 inches. Northern Florida and adjacent regions of Georgia and Alabama.

27a Pharyngeal teeth 1,4-4,1 to 4-4; a large round spot at base of caudal in young and juveniles; red postocular spot in life; numerous breeding tubercles in adults extending from posterior snout (internasal area) to the occiput, tubercles directed forward 28

27b Pharyngeal teeth 4-4 to 3-3; lacking round spot at base of caudal fin or with a small or faint spot; no red postocular spot in life; moderate to numerous breeding tubercles extending from tip of snout or internasal region to the occiput, tubercles erect 29

28a Pharyngeal teeth 1,4-4,1, occasionally a tooth missing in the outer row of one side; caudal peduncle scales usually 16 or 17; breeding tubercles or tubercle spots never present on the nape or on the sides of body. Fig. 214 . . . .
. . . . HORNYHEAD CHUB,
Nocomis biguttatus (Kirtland)

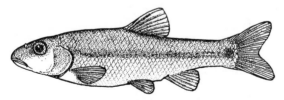

Figure 214.

Dark olive above, pale below. Young individuals have more distinct lateral band and caudal spot. Length 8-10 inches. Red River of Minnesota, North Dakota, and Manitoba, Canada. Wyoming and North Dakota to the Hudson River and south to northeastern Oklahoma and northern Ohio River drainage.

28b Pharyngeal teeth 4-4; caudal peduncle scales usually 19-20; adults with tubercles or tubercle spots on anterior portion of the nape and in rows on sides of the body . . . .
. . . . ORANGEFIN CHUB,
Nocomis effusus Lachner and Jenkins

Similar in coloration to hornyhead chub; medium sized, up to 8 inches in length. Central and western Ohio River basin, Cumberland, Green and lower Tennessee River drainages.

29a Intestine S-shaped, without an anterior ventral loop 30

29b Intestine with a ventral anterior loop covering portion of digestive tract . . . . . .
. . . . BLUEHEAD CHUB,
Nocomis leptocephalus (Girard)

Stout bodied fish, up to 8 inches total length; with fewer than 25 breeding tubercles; head bluish in breeding males. Potomac River south to Santee River drainage of Atlantic Coast; upper New River of Virginia.

30a Scales large, scales in lateral line average less than 40; circumferential scales average less than 32. Fig. 215. . . . .
. . . . RIVER CHUB,
Nocomis micropogon (Cope)

Figure 215.

Dark olive above, pale below. Length 6 to 9 inches. Wabash River to Michigan and western New York and south to Virginia and on west side of Appalachians to northern Georgia and Alabama.

30b Scales smaller; scales in lateral line average more than 40; circumferential scales average more than 32 . . . .
. . . . **BULL CHUB,** *Nocomis raneyi* **Lachner and Jenkins**

Blackish-olive colored minnow, up to 10 inches in total length; faint or small basicaudal spot; horizontal lateral band present; breast partially scaled or naked. James, Chowan, and Roanoke drainages of Virginia; Tar and Neuse Rivers in eastern North Carolina.

BIGMOUTH CHUB, *Nocomis platyrhynchus* Lachner and Jenkins. Similar to the bull chub, but with wider mouth, always greater than 9.5 percent of standard length; breast usually well scaled. New River drainage of West Virginia, Virginia, and North Carolina.

31a Body irregularly speckled with black spots or small blotches, or with 8 to 11 blotches on a dusky lateral band 32

31b Body not speckled and with no blotches along the lateral line. 37

32a Upper lobe of caudal fin light, lower lobe with dark pigment and light margin . . . 33

32b Both lobes of caudal fin light or only faintly pigmented at the base 34

33a Scales of lateral line 40-43; dorsal fin not pointed; sides speckled; each scale above lateral line with a strong ridge. Fig. 216 . .
. . . . STURGEON CHUB, *Hybopsis gelida* (Girard)

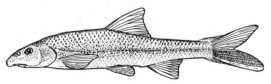

Figure 216.

Dusky above and silvery below; barbel short but distinct; eye small. Teeth 4-4. Length about 3 inches. Missouri River drainage.

33b Scales of lateral line 46-50; dorsal fin very pointed; sides not speckled; scales without ridges. Fig. 217 . . . .
. . . . SICKLEFIN CHUB, *Hybopsis meeki* **Jordan and Evermann**

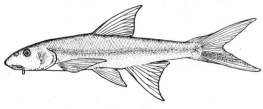

Figure 217.

Silvery with dusky lateral band and faint caudal spot. All fins very large, the pectoral reaching far past the base of the pelvic fins; barbel long; eye small. Teeth 4-4. Length 2 inches. Missouri River drainage.

34a Blotches present on sides along lateral band 35

34b Blotches not present on sides along lateral band 36

35a Scales in lateral line 43-51. Fig. 218 . . . . . STREAMLINE CHUB, *Hybopsis dissimilis* (Kirtland)

Figure 218.

Olivaceous above and light below with a pale bluish lateral band and a caudal spot; slightly speckled sometimes above lateral line; head rather broad. Teeth 4-4. Length 4 inches. Iowa and the Ohio River drainage north of the Ohio River to Oklahoma. Includes form called *H. wautaga* Jordan and Evermann in the upper Tennessee River drainage which is now considered the same species.

35b Scales in lateral line 40-43 . . . .
. . . . **BLOTCHED CHUB,**
Hybopsis insignis **Hubbs and Crowe**
Sides with row of dark blotches (large as pupil) and with scattered specks; mid-dorsal row of blotches present. Teeth 4-4. Length 3 inches. Cumberland and Tennessee River systems.

36a Barbel as long or longer than pupil of eye; sides of body rather heavily sprinkled with black dots. Fig. 219 . . . .
. . . . **SPECKLED CHUB,**
Hybopsis aestivalis **(Girard)**

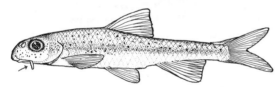

Figure 219.

Silvery and more or less heavily speckled; anal rays usually 8; belly naked. Teeth 4-4. Length 2 1/2 inches. Several sub-species ranging from upper Missouri River drainage to the Rio Grande and to western Florida. *H. a. tetranema* Gilbert with two pairs of barbels occurs in the Arkansas River drainage of Oklahoma, Arkansas, and Kansas.

36b Barbel shorter than pupil of eye; sides and back marked with scattered "X"-shaped spots. Fig. 220. . . . .
. . . . **GRAVEL CHUB,**
Hybopsis x-punctata **Hubbs and Crowe**

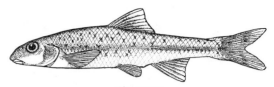

Figure 220.

Silvery with no distinct spots on pale lateral band; anal rays usually 7. Belly scaled. Teeth 4-4. Length about 4 inches. Southern Minnesota to Ohio and to Oklahoma.

37a Lateral line scales usually more than 50; occasionally some individuals may have as few as 48 . . . . **38**

37b Lateral line scales less than 50, usually less than 48 . . . . **39**

38a Dorsal fin distinctly falcate; no spot at base of caudal fin. Fig. 221 . . . .
. . . . **FLATHEAD CHUB,**
Hybopsis gracilis **(Richardson)**

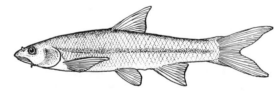

Figure 221.

Pale olive above and silvery below; barbel rather long; head broad, width equal to depth; mouth large, reaching past front of eye. Teeth 2,4-4,2. Length up to 12 inches. Several sub-species in streams of the plains from Yukon Territory south to Oklahoma.

38b Dorsal fin not falcate; a dark caudal spot present. Fig. 222 . . . .
. . . . **SPOTFIN CHUB,**
Hybopsis monacha **(Cope)**

Figure 222.

Light olive above, silvery below; black blotch in upper part of dorsal fin; barbel small; head not broad; mouth small, not reaching past eye. Teeth 4-4. Length about 4 inches. Upper Tennessee River drainage.

39a Dorsal fin with spot in upper posterior portion. Fig. 223 . . . .
. . . . **THICKLIP CHUB,**
Hybopsis labrosa (Cope)

Figure 223.

Snout long and blunt; barbels very long; small spot at base of caudal fin. Males are steel blue with black markings on back, especially at base of caudal fin. Females are pale silvery with bluish streak on caudal peduncle. Teeth 1,4-4,1. Length 3 inches. Santee River drainage, North and South Carolina.

39b Dorsal fin without spot **40**

40a Origin of dorsal fin above or behind origin of pelvic fin: pigment present at base of anal fin **41**

40b Origin of dorsal fin anterior to origin of pelvic fin no pigment present at base of anal fin **44**

41a Caudal spot present; light streak a scale or two wide above the lateral band **42**

41b Caudal spot faint or absent; no light streak above the lateral band **43**

42a Barbel minute; pharyngeal teeth 4-4 . . . .
. . . . **REDEYE CHUB,**
Hybopsis harperi (Fowler)
(See Fig. 213 and couplet 26.) Individuals without a sub-terminal mouth will key here.

42b Barbel well developed; pharyngeal teeth usually 1,4-4,1. Fig. 224 . . . .
. . . . **LINED CHUB,**
Hybopsis lineapunctata Clemmer and Suttkus

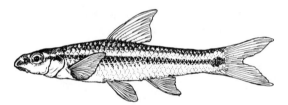

Figure 224.

43a Pharyngeal teeth 1,4-4,1. Fig. 225 . . . . . .
. . . . **BIGEYE CHUB,**
Hybopsis amblops (Rafinesque)

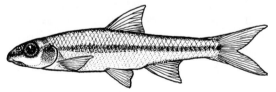

Figure 225.

Dusky green above and silvery below with a prominent silvery lateral band which is often heavily pigmented. Length 2 to 3 inches. Western New York and northwestern Pennsylvania, Ohio, southeastern Michigan to Indiana, Kentucky, West Virginia, Tennessee,

Missouri, northern Arkansas and Tennessee River drainage in Alabama.

ROSYFACE CHUB, *Hybopsis rubrifrons* (Jordan). Similar to bigeye chub except the head is longer and narrower. Altamaha River system in Georgia, Savannaha River, and Santee River, South Carolina.

CLEAR CHUB, *Hybopsis winchelli* Girard. Similar to the preceding bigeye and rosyface chub. Pigmentation variable, prominent dark lateral band extending through eye and around snout. Anal rays 8. Length slightly less than 3 inches. Flint River of Georgia and Florida to Lake Pontchartrain drainage of Louisiana.

43b Pharyngeal teeth in single row 4-4 . . . . . .
. . . . SLENDER CHUB,
Hybopsis cahni **Hubbs and Crowe**
Dusky above, light below, caudal spot may be faint. Body slender and with long slender peduncle. Length 2 1/2 inches. Powell and Clinch Rivers, Tennessee.

44a Eye small, 3 3/4 times in head length;
dorsal fin not falcate. Fig. 226 . . . .
. . . . HIGHBACK CHUB,
Hybopsis hypsinota **(Cope)**

Figure 226.

Silvery with dusky lateral band passing around snout; faint caudal spot present. Males are deep violet luster with pink fins. Teeth 1,4-4,1. Length 3 inches. Santee River drainage, North and South Carolina.

44b Eye large, 3 to 3 1/3 times in head length;
dorsal fin falcate. Fig. 227 . . . .
. . . . SILVER CHUB,
Hybopsis storeriana **(Kirtland)**

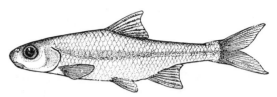

Figure 227.

Pale greenish above and silvery below with a slight dusky band and no caudal spot. Teeth 1,4-4,1. Length up to 10 inches. Red River drainage of North Dakota to New York and south to northern Alabama and Oklahoma.

45a Lateral line scales in body length more
than 45 46

45b Lateral line scales in body length less than
45 70

46a Lateral line incomplete or undeveloped . .
. . . . . *Chrosomus*
SOUTHERN REDBELLY DACE, *Chrosomus erythrogaster* (Rafinesque). Fig. 228. Olive

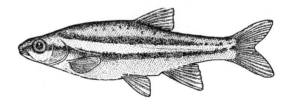

Figure 228.

brown to blackish above and light below; two lateral dark bands with reddish streak between; scale rows about 85. Males with reddish bellies. Mouth does not reach to eye (Fig. 229). Length about 3 inches. Southern Minnesota to Pennsylvania and south to northern Alabama, Arkansas, and Oklahoma.

Figure 229.

NORTHERN REDBELLY DACE, *Chrosomus eos* Cope. Similar to southern redbelly dace except snout is shorter and mouth reaches to eye (Fig. 230). Length up to 3 inches. Northern British Columbia to Hudson Bay drainage and Nova Scotia and south to Montana, Colorado, central Minnesota and to Pennsylvania. Isolated population in Nebraska.

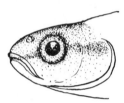

Figure 230.

MOUNTAIN REDBELLY DACE, *Chrosomus oreas* Cope. Fig. 231. Brightly colored with upper lateral band mostly posterior to anus. Length 3 inches. Upper James, Roanoke, and Kanawha Rivers.

Figure 231.

FINESCALE DACE, *Chrosomus neogaeus* (Cope). Fig. 232. Dark above and light below with dark lateral band above which is a light streak; more than 80 scale rows. Length up to 5 inches. Northwestern Canada to New Brunswick and New England and south to Montana, Colorado, northern Minnesota, Wisconsin, and Michigan. Isolated populations in Nebraska and Black Hills.

Figure 232.

46b   Lateral line complete **47**

47a   Pharyngeal teeth in one row, 4-5, 5-5, 5-4, or 6-6. . . . . **48**

47b   Pharyngeal teeth in 2 rows, 2,4(5)-(4)5,2. . . . . **52**

48a   Anal fin rays 10 or more **49**

48b   Anal fin rays usually less than 10 . . . . . **50**

49a   Belly behind pelvic fins with a sharp naked keel; length of intestine about twice length of body . . . . *Notemigonus*
GOLDEN SHINER, *Notemigonus crysoleucas* (Mitchill). Fig. 233. Body deep; silvery gold color; lateral line deeply decurved. Teeth 4-4 or 5-5. Length up to 10 inches. Many sub-species, Saskatchewan to Quebec and southward to Florida and south central Texas. Introduced west of Rockies.

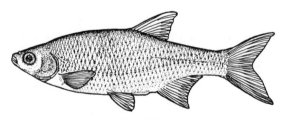

Figure 233.

49b Belly behind pelvic fins not with sharp naked keel; length of intestine more than twice length of body .... *Lavinia*

HITCH, *Lavinia exilicauda* Baird and Girard. Fig. 234. Deep bodied; dark above, light below; mouth short, not extending much behind nostrils; lateral line deeply decurved. Teeth 4(5)-5. Length up to 12 inches. Several sub-species in streams of central California.

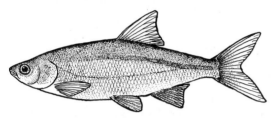

Figure 234.

50a Scales before dorsal fin 44-51; spot on base of caudal fin rays; sides more or less blotched.................... *Moapa*

MOAPA DACE, *Moapa coriacea* Hubbs and Miller. Fig. 235. Deep olive above with blotches on sides and white on belly; side marked with a golden brown lateral band above which is a light streak. Premaxillary is actually non-protractile, but the frenum connecting it is slight and hidden in the groove. Teeth usually 5-4. Resembles *Gila* and *Rhinichthys* but has only one row of pharyngeal teeth and lacks barbel. Length 2 inches. Moapa River, Nevada.

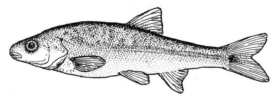

Figure 235.

50b Scales before dorsal fin less than 40; no distinct spot on base of caudal fin rays; sides not greatly blotched, if any **51**

51a Front of dorsal fin well behind front of pelvic fins; 32-38 scales before dorsal fin.................... *Hesperoleucas*

CALIFORNIA ROACH, *Hesperoleucas (Lavinia) symmetricus* (Baird and Girard). Fig. 236. Dusky above, pale below with a partial lateral band. Teeth 5-4. Length 5 inches. Several sub-species, central and northern California.

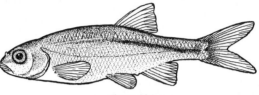

Figure 236.

51b Front of dorsal fin about over front of pelvic fins; 26-33 scales before dorsal fin . ... TUI CHUB, *Gila bicolor* (Girard)

Individuals with teeth 5-5 (in one row) will key here. (See Fig. 239, couplet 57.)

52a Upper sides more or less strongly blotched or mottled.............. *Rhinichthys*

Those species and individuals which lack a frenum or have one so minute it is overlooked and have poorly developed barbels or have no barbels will key here. (See couplet 12a, Figs. 183, 185, 186.)

52b Upper sides not blotched or mottled but may be dark..................... **53**

53a Snout long and rather pike-like; upper end of preopercle closer to end of opercle than to eye; pharyngeal teeth far apart and scarcely hooked .... *Ptychocheilus*

NORTHERN SQUAWFISH, *Ptychocheilus oregonensis* (Richardson). Fig. 237. Dusky green above, silvery below; lateral line scales

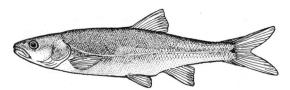

Figure 237.

67-75; 46-56 scales before dorsal fin; dorsal rays 9-10; anal rays 8. Teeth 2,4-5,2. Length up to 2 feet. Widespread in Columbia River drainage and in coastal streams of Oregon and Washington.

SACRAMENTO SQUAWFISH, *Ptychocheilus grandis* (Ayres). Fig. 238. Dark above, light below; lateral line scales 90-95; 36-41 scales before dorsal fin; dorsal rays 9; anal rays 7. Teeth, 2,5-4,2. Length 20 inches. Lower Sacramento River drainage, California.

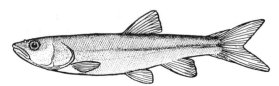

Figure 238.

UMPQUA SQUAWFISH, *Ptychocheilus umpquae* Snyder. Similar to northern squawfish but with smaller scales and only 8 anal rays. Sinslaw and Umpqua River drainages, Oregon.

COLORADO SQUAWFISH, *Ptychocheilus lucius* Girard. Dark above and light below; lateral line scales 83-87; dorsal rays 9; anal rays 9. Teeth 2,4-5,2. Length nearly 5 feet; largest of all American minnows. Lower Colorado River drainage.

53b   **Snout shorter than in squawfish, not pike-like; upper end of opercle closer to eye than to end of opercle; pharyngeal teeth close together and strongly hooked** **54**

54a   **Scales crowded before the dorsal fin** **67**

54b   **Scales not crowded before dorsal fin, predorsal scales sometimes embedded or predorsal region only partially scaled** **55**

55a   **Origin of dorsal fin behind origin of pelvic fins** **56**

55b   **Origin of dorsal fin over origin of pelvic fins** **65**

56a   **Scales large, less than 70 in the lateral line** **57**

56b   **Scales small, more than 70 in the lateral line** **62**

57a   **Pharyngeal teeth in a single row, 5-5, 5-4, or 4-4. Fig. 239** . . . . **TUI CHUB, *Gila bicolor* (Girard)**

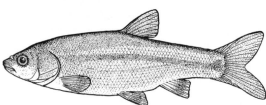

Figure 239.

Dusky brown to olive in color with a silvery belly, with a dusky lateral band. Scales in lateral line 40-60. Anal rays 7-9, usually 8, dorsal rays 7-9, usually 8. Owens and Mojave rivers, upper Pit River, introduced into several reservoirs in Sacramento River, California. Interior basins Oregon and Nevada, introduced into Columbia River system Washington and Oregon. Many sub-species.

57b Pharyngeal teeth in two rows, typically 2,5-4,2. 58

58a Scales in the lateral line usually less than 60 59

58b Scales in the lateral line usually more than 60 61

59a Anal fin rays 7. . . . . **ARROYO CHUB,** *Gila orcutti* **(Eigenmann and Eigenmann)**
Small fish, usually less than 4 inches in total length but occasionally reaching a length of 12 inches. Silvery to greyish green on back, with a dull grey lateral band and white belly. Scales in the lateral line 48-62. Pharyngeal teeth usually 2,5-4,2, but variable. Streams of the Los Angeles Plain, upper Santa Clara River, introduced into the Mojave, Santa Maria, and Cuyama river systems, California.

59b Anal fin rays 8, rarely 7 in *Gila purpurea* 60

60a Pelvic fin rays 9. . . . . **THICKTAIL CHUB,** *Gila crassicauda* **(Baird and Girard)**
Heavy bodied fish with a short, deep and thick caudal peduncle. Olivaceous above and pale below. Length 10-12 inches. Rare, possibly extinct. Streams tributary to San Francisco Bay, Clear Lake, and lowland areas of Central Valley, California.

60b Pelvic fin rays 8, rarely 7 . . . . **YAQUI CHUB,** *Gila purpurea* **(Girard)**
Dark fish, lacking lateral band, with small triangular spot at caudal base. Scales in the lateral line less than 59. Pharyngeal teeth 2,5-4,2. Rio Yaqui River, southeastern Arizona.

61a Dorsal fin rays 8. Fig. 240 . . . . **RIO GRANDE CHUB,** *Gila nigrescens* **(Girard)**

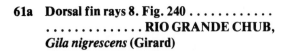

Figure 240.

Dusky above, silvery below. Terminal mouth; scales in lateral line 55-70, usually more than 65. Length of 5-6 inches in streams and 10 inches in lakes. Rio Grande and Pecos river drainages, headwaters of the Canadian River, Texas, and New Mexico.

PECOS CHUB, *Gila pandora* (Cope). Similar to the Rio Grande chub. Length of 8 inches. Scales in the lateral line 59-65. Pecos River basin in New Mexico and Texas.

61b Dorsal fin rays 9. Fig. 241 . . . . **BLUE CHUB,** *Gila coerulea* **(Girard)**

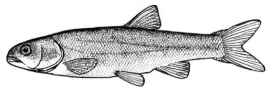

Figure 241.

Small bluish silver chub, seldom more than 14 inches total length. Anal rays 8 or 9. Scales in the lateral line 58-71. Pharyngeal teeth 2,5-4,2. Dorsal origin only slightly behind origin of pelvic fins. Klamath and Lost river systems of Oregon and California.

62a Dorsal and anal fin with 8 rays. Fig. 242 . . . . **LEATHERSIDE CHUB,** *Gila copei* **(Jordan and Gilbert)**

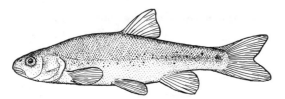

Figure 242.

Skin texture rather leathery. Body bluish above, silvery below with a dusky lateral streak; scales in the lateral line 75-85. Pharyngeal teeth (1)2,4-4,2(1). Length up to 6 inches. Bonneville Basin streams draining into Great Salt Lake and Utah Lake, Utah. Upper part of Snake River in Wyoming, Little Wood River, Idaho. Introduced into the Colorado River.

62b Dorsal and anal fins with more than 8 rays 63

63a Dorsal fin rays usually 9, anal fin rays usually 9 or less; caudal peduncle not pencil-like. Fig. 243 . . . . **ROUNDTAIL CHUB,** *Gila robusta* **Baird and Girard**

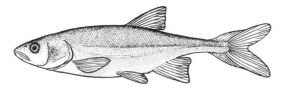

Figure 243.

Dusky to greenish on back and upper sides and pale below. Caudal peduncle depth divided into head length usually less than 4.0. Native to the Colorado, headwaters in Wyoming and Colorado, Utah, New Mexico, and Arizona into Mexico. At least four sub-species are recognized.

63b Dorsal fin rays 9 or 10, anal fin rays 10 or more; caudal peduncle very thin and pencil-like 64

64a Prominent hump on nape, sometimes overhanging the occiput; mouth inferior, overhung by fleshy snout. Fig. 244 . . . . **HUMPBACK CHUB,** *Gila cypha* **Miller**

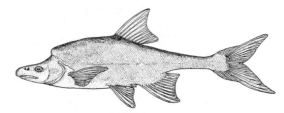

Figure 244.

The distinctly humped back serves to identify this chub. It reaches a length of 17 inches. Once common throughout rapid water in the Colorado, now found only in the Colorado River in the Grand Canyon and above, including the Green River of Utah.

64b Dorsal hump less prominent, hump rises smoothly from margin of occiput, never overhanging occiput; mouth terminal or slightly inferior, not overhung by fleshy snout. . . . . **BONYTAIL,** *Gila elegans* **Baird and Girard**

Greenish to bluish on the back and sides with a white belly. Length up to 20 inches. The nuchal hump less pronounced than that of the Humpback chub. The caudal fin is deeply forked. Scales in the lateral line, scales embedded, 75-88. Throughout the Colorado River and its larger tributaries.

65a Scales large, less than 65 in the lateral line. Fig. 245 . . . . **UTAH CHUB,** *Gila atraria* **(Girard)**

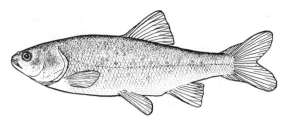

Figure 245.

Bluish to brassy colored chub with a silvery belly. Large, over 20 inches in total length. Dorsal fin rays 8-10, usually 9, anal fin rays 7-9, usually 8. Scales in the lateral line 45-65. Upper Snake River, basin of glacial Lake Bonneville in Utah, Idaho, Wyoming, and Nevada; upper Colorado River in Utah.

65b **Scales small, more than 65 in the lateral line** **66**

66a **Body with two prominent black bands, one on each side of the lateral line; basicaudal spot oval in shape.** . . . . **SONORA CHUB,** *Gila ditaenia* **Miller**
Small fish, usually less than 6 inches in total length. Scales in the lateral line 63-75. Dorsal fin rays 8, rarely 9, anal and pelvic fin rays 8. Bear Canyon, west Nogales, Santa Cruz County, Arizona.

66b **Color dark all over, sometimes belly lighter, without black bands; no basicaudal spot** . . . . **GILA CHUB,** *Gila intermedia* **(Girard)**
Similar to the Sonora chub in shape but differing in lacking prominent bands and basicaudal spot. Dorsal fin rays usually 8 or fewer, rarely 9, anal fin rays 8 or fewer. Central and southern Arizona in small creeks and artificial impoundments.

67a **Snout rather pointed; jaws extending to or behind front margin of eye** **68**

67b **Snout rather blunt; jaws not quite reaching the front margin of the eye.** . . **69**

68a **Scales in lateral line fewer than 60; caudal peduncle length equal to length of lower jaw. Fig. 246** . . . . **ROSYFACE DACE,** *Clinostomus funduloides* **Girard**

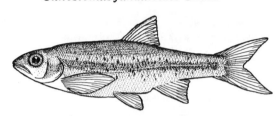

Figure 246.

Bluish green somewhat mottled above, pale below; reddish in breeding males. Scales in the lateral line 48-53; usually 9 dorsal rays; 8 anal rays. Teeth 2,5-5(4),2. Length to 5 inches. Headwaters of streams from Chesapeake Bay to North Carolina.

68b **Scales in the lateral line more than 60; caudal peduncle length less than length of lower jaw. Fig. 247** . . . . **REDSIDE DACE,** *Clinostomus elongatus* **(Kirtland)**

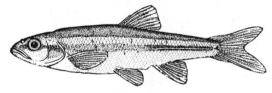

Figure 247.

Dusky blue above, sides and belly silvery, breeding males with reddish; lateral band present; 65-70 scales in the lateral line; dorsal rays usually 8; anal rays 9. Teeth 2,5-4,2.

Length to 5 inches. Southeastern Minnesota through part of the eastern Great Lakes and Ohio River drainages south to northeastern Oklahoma.

69a **Anal fin rays 10 or less, usually 9; dorsal fin rays usually 8 or less. Fig. 248.** . . . . **LAHONTAN REDSIDE,** *Richardsonius egregius* (Girard)

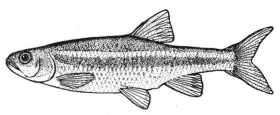

Figure 248.

Very dark on back with golden belly, red in breeding males; sides marked with 2 dark lateral bands with reddish streak between. Pharyngeal teeth 2,4-5,2. Length to 4 inches. Lahontan Basin in western Nevada and southeastern California.

69b **Anal fin rays more than 10, usually 15; dorsal fin rays 8 or more usually 10. Fig. 249.** . . . . **REDSIDE SHINER,** *Richardsonius balteatus* (Richardson)

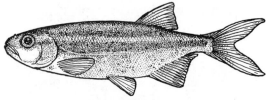

Figure 249.

Dark above with silvery sides and belly, rosy in breeding males; 9-11 dorsal rays; anal rays more than 10 (16+). Teeth 2,5-4,2. Length up to 6 inches. Columbia River drainage and Salt Lake Basin.

70a **First rudimentary dorsal ray is more or less thickened and distinctly separated from the first well-developed ray by a membrane (Fig. 250)** **71**

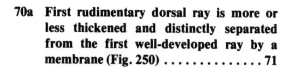

Figure 250.

70b **First rudimentary dorsal ray is a thin splint closely attached to the first long ray (Fig. 251)** **74**

Figure 251.

71a **Peritoneum black or greyish black; intestine long with one to several loops crossing the ventral mid-line** **72**

71b **Peritoneum silvery; intestine short, s-shaped** **73**

72a **Lateral line incomplete; mouth terminal and oblique; lateral band diffuse or absent; caudal spot diffuse or absent. Fig. 252.** . . . . **FATHEAD MINNOW,** *Pimephales promelas* Rafinesque

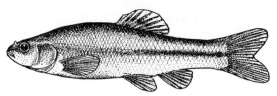

Figure 252.

Olivaceous in color. Adults have a horizontal dark bar across the dorsal fin nearly halfway up. Breeding males (Fig. 253) are very dark with vertical bands, tubercles on the snout, and a heavy pad on the nape and back anterior to the origin of the dorsal fin. Circumferential scales 38 or more. Teeth 4-4. Length to 2 1/2 inches. Several sub-species. Widespread from northwestern Canada east of Rockies to Maine and southward to the Susquehanna and to the Gulf States, a sub-species in New Mexico. Widely introduced west of the Rocky Mountains.

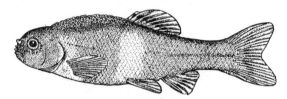

Figure 253.

72b Lateral line complete; mouth sub-terminal and horizontal, with overhanging snout; lateral band and caudal spot present. Fig. 254 . . . . BLUNTNOSE MINNOW, *Pimephales notatus* (Rafinesque)

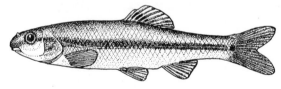

Figure 254.

Olivaceous in color. Body depth more than 4.5 times in standard length. Circumferential scales 32 or fewer. Length to 4 inches. Teeth 4-4. Widespread from North Dakota and Manitoba through southern Canada and the Great Lakes southward to Virginia, central Alabama and Oklahoma.

73a Dark lateral band present; caudal spot vertical. Fig. 255 . . . . BULLHEAD MINNOW, *Pimephales vigilax* (Baird and Girard)

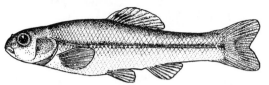

Figure 255.

Dusky yellowish above, silvery below with dark spot on the first four dorsal rays. Nuptial tubercles in two rows, 9 in number. Caudal peduncle usually more than 1/2 the caudal peduncle length. Teeth 4-4. Length to 3 inches. Several sub-species from southeastern Minnesota and West Virginia to northern Alabama and the Rio Grande Basin in Mexico.

73b Dark lateral band absent; caudal spot wedg haped. Fig. 256 . . . . SLIM MINNOW, *Pimephales tenellus* (Girard)

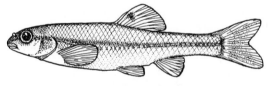

Figure 256.

Very similar and closely related to the bullhead minnow. Nuptial tubercles in three rows, 11-13 in number. Caudal peduncle depth usually less than 1/2 caudal peduncle length. Nesho River system of Oklahoma, Kansas and Missouri.

74a Dorsal rays usually 9 75

74b Dorsal rays usually 8 76

75a Mouth small and upturned, so it is almost vertical; scales cycloid, not scalloped on the posterior margin; pharyngeal teeth 0,5-5,0. Fig. 257 . . . .
. . . . PUGNOSE MINNOW, *Opsopoeodus emiliae* Hay

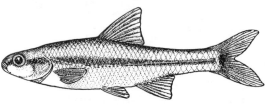

Figure 257.

Yellowish to silvery with a dark lateral band; breast scaleless. Anal rays 8. Scales in the lateral line 36-40. Relatively small minnow, up to 2 1/2 inches in length. Southeastern Minnesota to Michigan, south to Florida and Texas.

75b Mouth larger, oblique but not upturned and vertical; scales circular, scalloped on the posterior margin, exposed surface giving the appearance of being ctenoid; pharyngeal teeth (1),2,4-5,2 or 2,5-5,2. Fig. 258. . . . .
. . . . GRASS CARP, *Ctenopharyngodon idellus* Cuvier and Valenciennes.

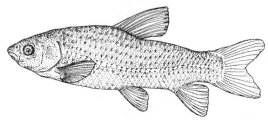

Figure 258.

Olivaceous to greenish brown with white ventral surface. Golden appearance when alive. Anal rays 9 or 10, usually 10. Scales in the lateral line 40-43. Introduced, native to China and southern Russia. In ponds in Arizona, Arkansas, and Iowa, in Florida waters.

Reported to have escaped into the Mississippi River.

76a Mandible, maxillary, sub-orbitals and sub-opercle with visible cavernous chambers. Fig. 259 . . . .
. . . . SILVERJAW MINNOW, *Ericymba buccata* Cope

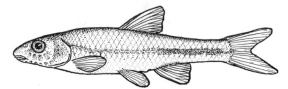

Figure 259.

Olivaceous above with silvery sides and belly. Teeth 1,4-4,1. Length to 5 inches. Distributed from southeastern Missouri to western Pennsylvania and south to Florida and Arkansas.

76b Mandible, maxillary, sub-orbitals and sub-opercles without visible cavernous chambers. 77

77a Lateral line absent or incomplete; inner row of teeth 4-5 or 5-4 78

77b Lateral line complete; inner row of teeth 4-4 79

78a Lateral line completely absent. Fig. 260. .
. . . . LEAST CHUB, *Iotichthys phlegethontis* (Cope)

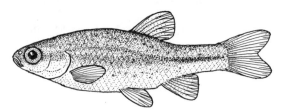

Figure 260.

Olivaceous with a broad lateral band and light below. Teeth 1,5-4,2. Small, about 1 1/2 inches long. Tributaries of Great Salt Lake and Lake Sevier, Utah.

78b Lateral line present but incomplete (about halfway). Fig. 261 . . . . FLAME CHUB, *Hemitrema flammea* (Jordan and Gilbert)

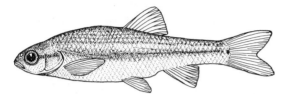

Figure 261.

Dark on back, pale below with dark lateral band above which is a light streak. Breeding males with red belly. Teeth 2,14-5,2. Length 2 /2 inches. Tributaries of upper Tennessee River in Tennessee and northern Alabama, Coosa drinage, Alabama and Georgia.

79a Intestine short "S"-shaped, less than twice the standard length; peritoneum usually silvery or speckled . . . . (p. 87) SHINERS, *Notropis*
This genus contains more species than any other genus of freshwater fishes in North America. One species, the Cape Fear Shiner, *Notropis mekistocholas* Snelson, has a long intestine.

79b Intestine longer, coiled more or less spring-like and more than twice the standard length; peritoneum black 80

80a Body with a prominent dark lateral band 81

80b Body silvery or yellowish without a prominent dark lateral band 82

81a Mouth U-shaped; dark lateral band extends around snout . . . . *Dionda*
OZARK MINNOW, *Dionda nubila* (Forbes). Fig. 262. Body dark, with dark lateral band around snout, through eye, and on the sides to a faint caudal spot. Teeth 4-4. Length 2 1/2 inches. Wyoming to Illinois and south to the Ozarks.

Figure 262.

ROUNDNOSE MINNOW, *Dionda episcopa* Girard. Fig. 263. Similar to the Ozark minnow except the snout is blunter and the mouth is smaller. Several sub-species. Oklahoma through central Texas and northern New Mexico.

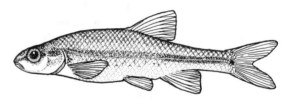

Figure 263.

DEVILS RIVER MINNOW, *Dionda diaboli* Hubbs and Brown. Similar to the roundnose minnow but with a wedge-shaped caudal spot. Devil's River and nearby tributaries of the Rio Grande River, Texas.

81b Mouth C-shaped; dark lateral band not extending around the snout. Fig. 352 . . . . (see couplet 61a, p. 108)
CAPE FEAR SHINER, *Notropis mekistocholas* Snelson

82a Color yellowish; scales with about 20 faint radii. Fig. 264, 265 . . . .

. . . . **BRASSY MINNOW,**
Hybognathus hankinsoni **Hubbs**

Figure 264.

Figure 265.

Color brassy or yellowish in life. Head blunt. Length to 4 inches. Ranges from Montana to Lake Champlain and southward to Nebraska, Missouri and Colorado.

82b Color silvery; scales with fewer than 20 radii, radii usually prominent. Fig. 266 83

Figure 266.

**83a Edges of scales heavily pigmented, giving a diamond appearance; eye in adult as long as snout. Fig. 267 . . . .
. . . . CYPRESS MINNOW,**
Hybognathus hayi **Jordan**

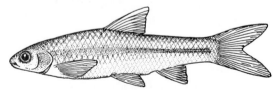

Figure 267.

Silvery with a dark back and hint of lateral band. Similar to the silvery minnow but body more slender and with shorter snout. Length to 4 inches. Southern Indiana southward to Tennessee, Mississippi and Louisiana.

**83b Edges of scales not heavily pigmented, without crosshatched appearance; eye in adult less than snout length. Fig. 268 . . . .
. . . . SILVERY MINNOW,**
Hybognathus nuchalis **Agassiz**

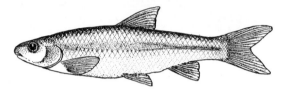

Figure 268.

A silvery minnow. Eye large, usually less than 5 times in length of head. Scales rows across the belly just anterior to pelvic fin 12-15, not including lateral line scales. Posterior process of the basioccipital bone broad. Scales in lateral line 33-41. Length to 6 inches. Several subspecies ranging from the Missouri drainage of Montana, southeastern Minnesota to Lake Champlain and south to the Gulf, absent from the Great Lakes drainage.

PLAINS MINNOW, *Hybognathus placitus* Girard. Fig. 269. Silvery in color. Eye smaller, usually 5 or more times in length of head. Scales rows around belly just anterior to pelvic fin 16-18, not including the lateral line scales. Posterior end of basioccipital process narrow, rod-like. Length up to 5 inches. Upper Missouri drainage and southward west of the Mississippi River from Montana, Wyoming,

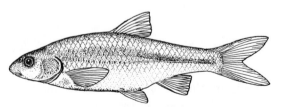

Figure 269.

the Dakotas to Arkansas, Texas and New Mexico.

84a **Scales present although minute; desert minnows** . . . . *Lepidomeda*
Apparently a complex of similar forms isolated in at least four desert stream and spring systems of Nevada, Utah and Arizona. Olivaceous above, silvery below with lateral band. Posterior dorsal spine in groove of anterior spine; first dorsal spine not sharp. Teeth 2,4-4,2. Length 3 inches.
    PAHRANAGAT SPINEDACE, *Lepidomeda altivelis* Miller and Hubbs. Two localities in Pahranagat Valley, Lincoln Co., Nevada. Now considered extinct.
    LITTLE COLORADO SPINEDACE, *Lepidomeda vittata* Cope. Headwaters of Little Colorado River system, Arizona.
    WHITE RIVER SPINEDACE, *Lepidomeda albivallis* Miller and Hubbs. White River system in eastern Nevada.
MIDDLE COLORADO SPINEDACE, *Lepidomeda mollispinis* Miller and Hubbs. Fig. 270. Santa Clara River, Virgin River system, Washington Co., Nevada, Utah and Arizona.

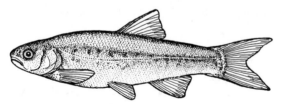

Figure 270.

84b **Scales absent; desert minnows** **85**

85a **Barbel present on maxillary** . . . .
. . . . *Plagopterus*
WOUNDFIN, *Plagopterus argentissimus* Cope. Fig. 271. Dusky back, silvery below. Teeth 2,5-4,2. Length 2 1/2 inches. Gila River, Arizona. Known at present only from the Virgin River system of Arizona, Utah and Nevada.

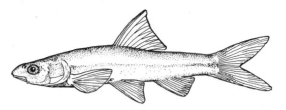

Figure 271.

85b **Barbel absent from maxillary** . . . . . *Meda*
SPIKEDACE, *Meda fulgida* Girard. Fig. 272. Dusky above, silvery on sides and below; somewhat speckled. Teeth 2,5-5,2. Length 3 inches. Gila River, Arizona and western New Mexico, probably several forms.

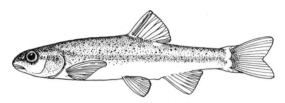

Figure 272.

SHINERS *GENUS NOTROPIS*

1a **Body depth usually less than 5 times in length from snout to base of caudal fin*** **2**

1b **Body depth usually 5 times or more in length from snout to base of caudal fin** **14**

2a **Body depth usually less than 4 times in length from snout to base of caudal fin** **3**

2b **Body depth usually 4 to 5 times in length from snout to base of caudal fin** **31**

*This is an unorthodox method for dividing this genus. Immature individuals and some sub-species may have greater or less body depth than that for adults or that characteristic for the species. Consequently, some species are keyed in several categories and may fall into quite unrelated groups.

3a Distinct spot present at base of caudal fin 4

3b Spot not present or not distinct at base of caudal fin 9

4a Spot at base of caudal fin large (approximately the size of the eye) and very black 5

4b Spot at base of caudal fin small (usually much smaller than the eye) or rather faint 6

5a Anal rays 7 or 8. Fig. 273. . . . . BLACKTAIL SHINER, *Notropis venustus* (Girard)

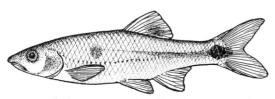

Figure 273.

Olivaceous to steel blue, slight dusky band. Somewhat variable throughout its range. Scales in lateral line 36-43. Pharyngeal teeth 1,4-4,1. Length to 5 inches. Southern Illinois and eastern Missouri south to the Gulf coast from the Rio Grande in Texas to the Suwanee River, Georgia; Red River in Texas and Oklahoma.

5b Anal rays usually 9 to 11. Fig. 274 . . . . BANDFIN SHINER, *Notropis zonistius* (Jordan)

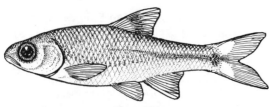

Figure 274.

Steel blue above, pale below; caudal fin with light spots at base; dorsal fin with horizontal bar; dark bar behind opercle. Teeth 2,4-4,2. Length 4 inches. Chattahoochee River drainage, Alabama, Georgia and Florida; upper Savannah River, Georgia.

6a Anal rays 7 to 8, usually 7 7

6b Anal rays 9 to 11 8

7a Scales before dorsal fin 12; maxillary reaching past eye; lower jaw slightly included. Fig. 275. . . . . TAMAULIPAS SHINER, *Notropis braytoni* Jordan and Evermann

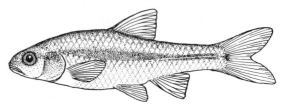

Figure 275.

Brownish with silvery lateral band. Teeth 4-4. Rio Grande drainage, Texas into Mexico.

7b Scales before dorsal fin 15-16; maxillary reaching to front of eye; lower jaw equal to upper. Fig. 276 . . . . TOPEKA SHINER, *Notropis topeka* Gilbert

Figure 276.

Olivaceous above with a dusky lateral band, belly white. Wedge-shaped spot at caudal base. Anal fin rays usually 7. Lateral line scales

32-36. Teeth 4-4. South Dakota and southern Minnesota to Kansas and Missouri.

OHOOPEE SHINER, *Notropis leedsi* Fowler. Fig. 277. Bluish above; silvery below with broad dark lateral band ending in faint caudal spot; adults with blotch halfway up on front of dorsal fin. Teeth 4-4. Length 2 3/4 inches. Savannah River to Ochlocknee River in Georgia and Florida.

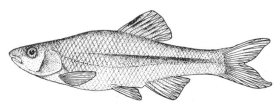

Figure 277.

8a   Basal part of caudal fin with pair of light spots. (See Fig. 274, couplet 5b.) . . . . BANDFIN SHINER, *Notropis zonistius* **(Jordan)**

Dorsal fin about over pelvic fin. Young have smaller caudal spots and will key here.

HIGHFIN SHINER, *Notropis altipinnis* (Cope). Fig. 278. Front of dorsal fin behind front of pelvic fin. Teeth 2,4-4,2. Many subspecies in coastal streams, Virginia to South Carolina.

Figure 278.

SAILFIN SHINER, *Notropis hypselopterus* (Gunther). Fig. 279. Front of dorsal fin far

Figure 279.

behind front of pelvic fins; dorsal fin with blotch; caudal fin with 2 rosy spots at base; sides with wide silvery lateral band. Teeth 1,4-4,1. Length 2 1/2 inches. Western Florida into Mississippi.

BROADSTRIPE SHINER, *Notropis euryzonus* Suttkus. Similar to sailfin shiner, but differs in having a more rectangular dorsal fin (males) and in pigmentation. Lateral band wide and caudal fin with small clear area in center. Teeth 2,4-4,2. Length 2 1/4 inches. Chattahoochee River, Georgia and Alabama.

FLAGFIN SHINER, *Notropis signipinnis* Bailey and Suttkus. Fig. 280. Silvery with a very broad lateral band; base of caudal fin with 2 yellowish spots. Anal fin rays 9-12. Teeth 2,4-4,2. Length 2 1/2 inches. Gulf drainage of southeastern Louisiana to western Florida.

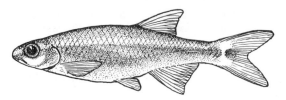

Figure 280.

8b   Basal part of caudal fin without a pair of light spots; fin marked with crescent-shaped bands. Fig. 281 . . . . FIREYBLACK SHINER, *Notropis pyrrhomelas* **(Cope)**

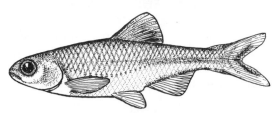

Figure 281.

Steel blue above and white below with scarlet bands on fins and head. Teeth 1,4-4,1. Length 3 1/4 inches. Santee and Pedee River drainages in North and South Carolina.

ALTAMAHA SHINER, *Notropis*

xaenurus (Jordan) May key here but caudal fin without crescent band. (See couplet 41a.)

9a First dorsal ray longer than last ray when fin is depressed. Fig. 282 . . . . **COMMON SHINER,** *Notropis cornutus* **(Mitchill)**

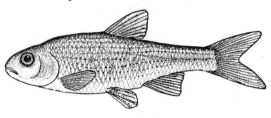

Figure 282.

Silvery with dusky dorsal stripe; no lateral band. Adults appear to be rather loosely scaled. Lateral line scales elevated except the last 8 or 10. Origin of dorsal fin slightly before origin of pelvic fins. Teeth 2,4-4,2. Length 7-8 inches. Southeastern Saskatchewan and Colorado eastward to the Appalachians.

CENTRAL COMMON SHINER, *Notropis c. chrysocephalus* (Rafinesque) with less than 22 rows of scales before the dorsal fin instead of more than 25 has recently been considered as a different species. Southern Great Lakes and St. Lawrence drainages south to Oklahoma, northern Alabama and Georgia.

CRESCENT SHINER, *Notropis cerasinus* (Cope). Fig. 283. Similar to the common shiner. Olive brown in color with black crescent-shaped marks on the side of body. Caudal peduncle scales usually 14 (13-15). Teeth 2,4-4,2. Length to 5 inches. Upper

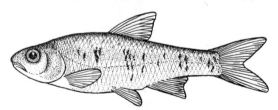

Figure 283.

Roanoke River, Virginia and North Carolina, eastern tributaries of the New River, Virginia.

WHITE SHINER, *Notropis albeolus* Jordan. Fig. 284. Similar to common shiner except snout is sharper. Last 18-20 lateral line scales not elevated. Teeth 2,4-4,2. Length 7 inches. Coastal drainage of Roanoke River in Virginia southward and on west side of divide in West Virginia.

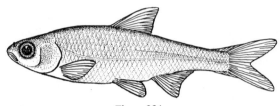

Figure 284.

PRETTY SHINER, *Notropis bellus* (Hay). Fig. 285. Dusky silvery with dusky lateral band and with orange on belly and fin bases. Readily distinguished by deeply pigmented margins of fins. Teeth 2,4-4,2. Length 2 1/2 inches. Gulf drainage of Alabama and Mississippi, Tennessee River system, Alabama.

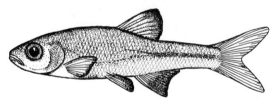

Figure 285.

9b First dorsal ray shorter than last ray when dorsal fin is depressed **10**

10a Dorsal fin without blotch but may be dusky. Fig. 286 . . . . **RED SHINER,** *Notropis lutrensis* **(Baird and Girard)**

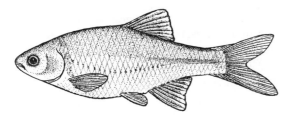

Figure 286.

Steel blue above and silvery below. Males with orange on fins and behind shoulders. Anal rays 8 or 9, rarely 10. Teeth 4-4 or 1,4-4,1. Wyoming to southern Minnesota and Illinois and southward to Louisiana and Mexico. Introduced into lower Colorado River, Arizona and California.

BEAUTIFUL SHINER, *Notropis formosus* (Girard). Very closely related to the red shiner except the anal fin has only 8 rays. Northern Mexico into southern New Mexico.

10b Dorsal fin with blotch or pigmentation on either anterior or posterior base 11

11a Anal rays usually 8. Fig. 287 . . . . SPOTFIN SHINER, *Notropis spilopterus* (Cope)

Figure 287.

Silvery with dark blotch on posterior part of dorsal fin. Teeth 1,4-4,1. Length 4 inches. Red River, North Dakota and Minnesota east through southern Ontario, Lake Champlain and the Potomac River, south to eastern Oklahoma, northern Arkansas then east to the Tennessee River in Alabama.

11b Anal rays usually 9 to 13 12

12a Dark blotch at base of anterior dorsal rays. Fig. 288. . . . . REDFIN SHINER, *Notropis umbratilis* (Girard)

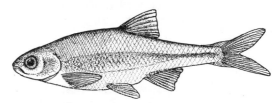

Figure 288.

Steel blue above and silvery below. Usually deep bodied, but depth varies with age and sub-species. Predorsal scale rows about 30. Teeth 2,4-4,2. Scales in the lateral line 41-47. Length to 3 inches. Southern Minnesota eastward through southern Ontario to Lake Ontario, in parts of Kentucky and West Virginia, south through Kansas to Mississippi, Louisiana and eastern Texas.

12b No blotch at base of anterior rays, but a dark blotch between last dorsal rays. . . 13

13a Caudal fin with pair of light spots at base of caudal fin. Fig. 289 . . . . BLUNTFACE SHINER, *Notropis camurus* (Jordan and Meek)

Figure 289.

Silvery with light area on base of caudal fin as in whitetail shiner (Fig. 326). Teeth 1,4-4,1.

Length 4 inches. Drainages of Arkansas, White, and Osage Rivers in Arkansas, northeastern Oklahoma, Kansas, Missouri, to Mississippi and Tennessee.

13b Caudal fin with no light areas on base. Fig. 290 . . . . SATINFIN SHINER,
Notropis analostanus **(Girard)**

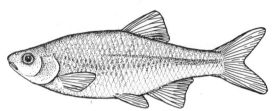

Figure 290.

Silvery color; lateral line scales 34-35. Teeth 1,4-4,1. Length 4 inches. Coastal drainage from the St. Lawrence to North Carolina.

STEELCOLOR SHINER, *Notropis whipplei* (Girard). Bluish gray to silver in color, similar to the spotfin shiner. Anal fin rays usually 9. Scales in the lateral line 36-39. Teeth 1,4-4,1. Widely distributed, Mississippi River basin from New York west to Indiana, south to Alabama, extreme northern Louisiana, Oklahoma, Arkansas and into Texas.

GREENFIN SHINER, *Notropis chloristius* (Jordan and Brayton). Black blotch in dorsal fin, fins pale at tips but no light area on base of caudal fin. Teeth 1,4-4,1. Sometimes confused with whitefin shiner. Santee River system, North and South Carolina. (See couplet 42a.)

TRICOLOR SHINER, *Notropis trichroistius* (Jordan and Gilbert). Fig. 291.

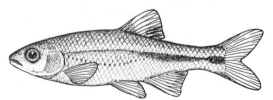

Figure 291.

Olivaceous with well-developed lateral band ending in elongated caudal spot. Dorsal fin dusky at base in front and on the last rays. Teeth 1, 4-4,1. Length 2 1/2 inches. Alabama River system.

14a Snout extending in a distinct bulge beyond upper lip 15

14b Snout not extending any appreciable distance before upper lip 17

15a Mouth distinctly oblique; spot present before base of caudal fin. Fig. 292 . . . . MIRROR SHINER,
Notropis spectrunculus **(Cope)**

Figure 292.

Olivaceous above, silvery below with dusky band, fins dusky. Teeth 4-4. Length 3 inches. Headwaters of Tennessee River.

15b Mouth not very oblique, but almost horizontal; no spot before base of caudal fin 16

16a Lateral band dusky, anal fin rays usually 8. Fig. 293 . . . . OZARK SHINER,
Notropis ozarcanus **Meek**

Figure 293.

Pale straw color on back, belly silvery white; with a dusky lateral band. Small blotch at base of first 2 or 3 dorsal fin rays. Anal fin rays usually 8. Scales in the lateral line 34-38. Teeth 4-4. Ozark Uplands, southern Missouri and northern Arkansas.

16b Lateral band faint, anal fin rays usually 7. Fig. 294 . . . . LONGNOSE SHINER, *Notropis longirostris* **(Hay)**

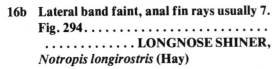

Figure 294.

Pale straw to sandy color with very faint lateral band, ventral surface white. Fins dusky, yellow in breeding individuals. Scales in the lateral line 34-37. Teeth 1,4-4,1(0). Length 2 1/2 inches. Gulf coast streams from eastern Louisiana to southeastern Georgia and western Florida.

17a Anal rays 7 to 8 18

17b Anal rays usually 9 to 11 23

18a Body depth scarcely over 5 (5.0 to about 5.2) times in body length 19

18b Body depth considerably over 5 (5.3 to over 6.0) times in body length 21

19a Lateral line very incomplete. Fig. 295 . . . . BRIDLE SHINER, *Notropis bifrenatus* **(Cope)**

Figure 295.

Straw-colored dorsum with a very black lateral band touching snout and chin, white on ventral surface. Teeth 4-4. Length 2 inches. Southern Maine south to Virginia in the Atlantic drainage; Lake Ontario and Champlain drainages.

19b Lateral line usually complete 20

20a Scales before dorsal fin 10 to 12. Fig. 296 . . . . MIMIC SHINER, *Notropis volucellus* **(Cope)**

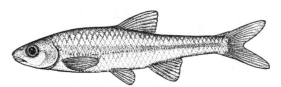

Figure 296.

Olivaceous with a very faint lateral streak, no spot before base of caudal fin. Snout U-shaped when viewed from above. Black pigment around anus. Scales elevated, 35-40 in the lateral line. Teeth 4-4. Length 2 1/4 inches. Southern half of Manitoba Ontario, southeastern Canada to Lake Erie, southward through Minnesota in the Mississippi River drainage to the Gulf coast, Mobile drainage westward to the Guadalupe River. Atlantic coast south of the Jame River to the Cape Fear River.

SWALLOWTAIL SHINER, *Notropis procne* (Cope). Fig. 297. Olivaceous with dark lateral band; faint spot sometimes before caudal fin; anal rays always 7. Teeth 4-4. Length 2 1/2 inches. Delaware River to South Carolina.

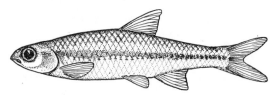

Figure 297.

**20b Scales before dorsal fin 14 to 18. Fig. 298
. . . . REDLIP SHINER,
Notropis chiliticus (Cope)**

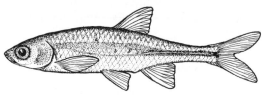

Figure 298.

Olivaceous with silvery lateral band. Males with red on head, especially on lips and on dorsal and anal fins. Teeth 2,4-4,2. Length 2 inches. Virginia and South Carolina.

HIGHSCALE SHINER, *Notropis hypsilepis* Suttkus and Raney. Lightly color, silvery. Resembles Coosa shiner but has black areas at base of first 4 or 5 dorsal and anal rays. Very pale lateral band. Small wedge-shaped caudal spot separate from the lateral stripe. Teeth 2,4-4,2. Length 2 inches. Appalachicola system, Alabama and Georgia.

COOSA SHINER, *Notropis xaenocephalus* (Jordan). Fig. 299. Silvery with dark lateral band ending in a confluent caudal spot; lateral band passes through eye and touches chin. Teeth 2,4-4,2. Length 2 inches. Alabama River system, Alabama and Georgia.

Figure 299.

GREENHEAD SHINER, *Notropis chlorocephalus* (Cope). Fig. 300. Olivaceous dusted with black specks and dusky lateral band. Males with much red. Teeth 2,4-4,2. Length 2 1/2 inches. North and South Carolina.

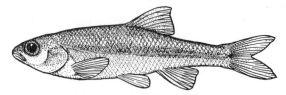

Figure 300.

WEED SHINER, *Notropis texanus* (Girard). Fig. 301. Slender weed shiners will key here. (See couplet No. 59a.)

Figure 301.

**21a Dorsal fin heavily pigmented and without spots or bars. Fig. 302 . . . .
. . . . TENNESSEE SHINER,
Notropis leuciodus (Cope)**

Figure 302.

Olivaceous, with silvery dark lateral band; caudal spot continuous with lateral band. Teeth (1)2,4-4,2(1). Length 3 inches. Upper Tennessee River drainage.

BURRHEAD SHINER, *Notropis asperifrons* Suttkus and Raney. Fig. 303. Silvery with a distinct lateral band and caudal spot. Teeth 2,4-4,2. Length 2 1/2 inches. Alabama and Black Warrior River systems, Alabama and Georgia.

Figure 303.

21b Dorsal fin with slight but distinct pigmentation **22**

22a Lateral line more or less incomplete. Fig. 304 . . . . **TAILLIGHT SHINER,** *Notropis maculatus* **(Hay)**

Figure 304.

Straw color, sometimes with reddish tinge and with very dark lateral and dorsal bands; dark streak on each side of anal base. Teeth 4-4. Length 2 1/2 inches. Mississippi River basin from western Kentucky, Missouri, Arkansas, Oklahoma, eastern Texas to Mississippi. Atlantic Coast from North Carolina to Florida.

BLUENOSE SHINER, *Notropis welaka* Evermann and Kendall. Similar to the taillight shiner. Exhibits marked sexual dimorphism, Male bluish with large flag-like dorsal fin, female light blue without large dorsal fin. Lateral line with 5 to 10 pored scales. Teeth 4-4 or 1,4-4,1. Length 2 1/4 inches. Eastern Florida west along the Gulf coast to Mississippi and the Pearl River, Louisiana.

22b Lateral line complete. (See Fig. 326, couplet 41a.) . . . .

. . . . **WHITEFIN SHINER,** *Notropis niveus* **(Cope)**
Slender immature individuals will key here.

23a Lateral line incomplete, 2/3 body length. Fig. 305 . . . . **RIBBON SHINER,** *Notropis fumeus* **Evermann**

Figure 305.

Yellowish-olive dorsum with silvery ventral surface. A diffuse lateral band present and some speckling. Anal fin rays 11-13. Teeth 2,4-4,2. Length 2 1/2 inches. Tennessee River basin to southeastern Missouri, Arkansas, eastern Oklahoma, Louisiana west to Texas.

23b Lateral line complete **24**

24a Scales in lateral line usually 40 or more. Fig. 306 . . . . **ROSEFIN SHINER,** *Notropis ardens* **(Cope)**

Figure 306.

Steel blue above and light below with dark lateral band. Teeth 2,4-4,2. Length 3 1/2 inches. Fins reddish in males. Several subspecies in Virginia (Roanoke River) and the upper and central Ohio River drainage.

CHERRYFIN SHINER, *Notropis roseipinnis* Hay. Fig. 307. Olivaceous above with a prominent mid-dorsal stripe, silvery below; slash-like black mark on the tips of 1st-3rd rays of dorsal and anal fins. Fins reddish in living specimens. Lateral line scales 43-49. Anal rays 11 sometimes 12. Teeth 2,4-4,2. Length to 2 1/4 inches. Coastal streams from Louisiana east of the Mississippi River to Alabama.

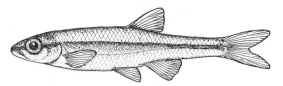

Figure 307.

MOUNTAIN SHINER, *Notropis lirus* Jordan. Fig. 308. Pale with dark lateral band through snout; spot in dorsal fin like that in *N. umbratilis*. Teeth 2,4-4,2. Length 2 1/2 inches. Coosa River system and upper Tennessee River, Alabama.

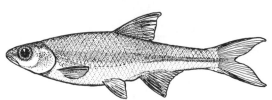

Figure 308.

24b Scales in lateral line usually less than 40 . . . . **25**

25a Front of dorsal fin over or only slightly before or behind front of pelvic fins. . . 26

25b Front of dorsal fin completely behind the middle of the base of the pelvic fins . . . 28

26a First dorsal ray longer than last dorsal ray when fin is depressed **27**

26b First dorsal ray equal or shorter than last dorsal when fin is depressed. Fig. 309 . . .
. . . . **SILVER SHINER,**
Notropis photogenus (Cope)

Figure 309.

Olivaceous, with a dark silvery lateral band. Anal fin rays 10-13. Scales in the lateral line 36-40. Teeth 2,4-4,2. Ohio River basin, Little Tennessee River, North Carolina to the Allegheny in New York, west to the Wabash in Indiana; Lake Erie drainage in Ohio and Indiana, Grand River, Ontario.

ROUGHHEAD SHINER, *Notropis semperasper* Gilbert. Resembles comely shiner but origin of dorsal fin is not as far behind the origin of the pelvic fins. Teeth 2,4-4,2. Upper drainage of the James River, Virginia.

27a Peritoneum silvery. Fig. 310 . . . .
. . . . **KIAMICHI SHINER,**
Notropis ortenburgeri Hubbs

Figure 310.

Pale with dusky lateral band extending through snout and chin. Anal fin rays 9 or 10. Teeth 4-4. Length 2 inches. Eastern Texas, southeastern Oklahoma and southwestern Arkansas.

COLORLESS SHINER, Notropis perpallidus Hubbs and Black. A slender very pale shiner. Scales large, usually 34 in the lateral line. Large round melanophores scattered over body, tiny melanophores faintly outline dorsal scales. Pigmented on top of head, lips and chin.

Teeth (1)2,4-4,2(1). Little River southeastern Oklahoma and southwestern Arkansas.

27b Peritoneum brownish or heavily covered with black melanophores. Fig. 311 . . . . TELESCOPE SHINER, *Notropis telescopus* **(Cope)**

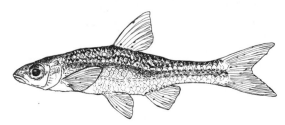

Figure 311.

Dorsal lateral part of back with 2 or 3 crooked stripes. Pigment above and below pores of lateral line scales, giving a stitch-like effect. Eye small, less than 1.5 times the length of snout. Anal rays usually 10 (9-12). Predorsal scales 13-15 (12-16). Teeth 2,4-4,2. Length 3 1/4 inches. Two disjunct populations; one west of the Mississippi River in Arkansas and Missouri, the other east of the Mississippi from Georgia and Kentucky eastward to Virginia and North Carolina.

POPEYE SHINER, *Notropis ariommus* (Cope). Scales dark edged above the lateral line, pale below; little or no pigment bordering lateral line scales. No stripes on dorsal lateral area. Eye larger, 1.5 times snout length. Anal rays usually 9 (8-10). Predorsal scales 15-18 (14-19). Teeth 2,4-4,2. White River in Arkansas and Missouri; Tennessee and Ohio Rivers in Indiana, Kentucky and Tennessee.

28a Length of snout greater than diameter of eye. Fig. 312 . . . . ROSYFACE SHINER, *Notropis rubellus* **(Agassiz)**

Figure 312.

Silvery with wide lateral band. Snout rather sharp. Anal fin rays 10-13. Scales in the lateral line 36-40. Teeth 2,4-4,2. Length 4 inches. North Dakota and Manitoba to the St. Lawrence and Hudson Rivers, south to Virginia and much of the Ohio River drainage.

28b Snout equal or less than diameter of eye. 29

29a Body slender and pale; head length 4.2 times or more in standard length. Fig. 313 . . . . EMERALD SHINER, *Notropis atherinoides* **Rafinesque**

Figure 313.

Pale silvery with a very faint lateral band. Head length 4.3-4.5 in standard length. Teeth 2,4-4,2. Length 4 inches. Common in deeper waters of lakes and rivers, occasionally entering mouths of tributary streams in the fall. Widely distributed from the Northwest Territories, Canada to Texas and Virginia.

SILVERSTRIPE SHINER, *Notropis stilbius* (Jordan). Pale green with silvery lateral band and with specks anteriorly and at base of caudal fin. Head length 4.2 times in standard length. Teeth 2,4-4,1. Length 3 inches. Gulf coast drainage of Mississippi, Alabama and Georgia.

29b Body heavier and darker above lateral line; head length usually less than 4.2 times in standard length 30

30a Less than 18 scales before dorsal fin . . . .
. . . . **POPEYE SHINER,**
Notropis ariommus (Cope)
Forms with front of dorsal fin far behind front of pelvics will key here. (See couplet 27b.)
COLORLESS SHINER, *Notropis perpallidus* Hubbs and Black. Individuals with front of dorsal fin far behind front of pelvic fins will key here. (See couplet 27a.)

30b More than 18 scales before dorsal fin. Fig. 314 . . . .
. . . . **COMELY SHINER,**
Notropis amoenus (Abbott)

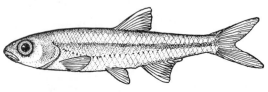

Figure 314.

Olive green slab-sided shiner, with faint lateral band; lower lip with pigment. Anal fin rays 10-12. Lateral line scales 36-42. Teeth (1)2,4-4,2(1). Length 4 inches. Atlantic coast drainage from southeastern New York, the Hudson River, New Jersey, eastern Pennsylvania to Cape Fear, North Carolina. Seneca Lake, in Finger Lakes-St. Lawrence drainage.
ROUGHHEAD SHINER, *Notropis semperasper* Gilbert. Similar to comely shiner but lacks pigment on lower lip and the front of the dorsal fin is not far behind the front of the pelvic fins. (See couplet 26b.)

31a Lateral line more or less incomplete (not reaching near base of caudal fin) 32

31b Lateral line usually complete (extending nearly to base of caudal fin) 34

32a Anal rays usually 7. (See Fig. 295, couplet 19a.) . . . .
. . . . **BRIDLE SHINER,**
Notropis bifrenatus (Cope)
Usually a slender minnow, but heavy individuals will key here.

32b Anal rays usually 8 or 9 33

33a Lateral band extends through eye and touches chin; dark edges of scales in row above and in lateral line form a zigzag pattern. Fig. 315 . . . .
. . . . **BLACKCHIN SHINER,**
Notropis heterodon (Cope)

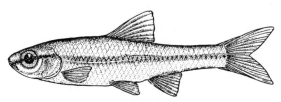

Figure 315.

Silvery. Lateral line incomplete with 34-38 scales. Teeth 4-4 to 1,4-4,1. Peritoneum silvery. Length 2 1/2 inches. North Dakota to southern Ontario and Quebec, south to New York and Iowa.

33b Lateral band extends through eye and around snout, but does not touch chin; dark edges of scales in lateral line, form a band of crescent marks; lateral line variable, may extend much more than 1/2 body length. Fig. 316 . . . .
. . . . **BLACKNOSE SHINER,**
Notropis heterolepis Eigenmann and Eigenmann

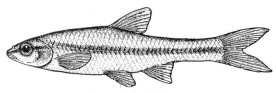

Figure 316.

Silvery. Lateral line almost complete, 34-38 scales. Teeth 4-4. Length 2 1/2 inches. Southern Canada from Saskatchewan to Nova Scotia, Maine and Ohio River drainage to Iowa in glacial lakes.

34a Mouth almost vertical and very small, extending only about halfway to the front of the eye. Fig. 317 . . . .
. . . . PUGNOSE SHINER,
Notropis anogenus Forbes

Figure 317.

Dark lateral band, chin black; faint caudal spot. Teeth 4-4. Scales in lateral line 34-38. Length 2 inches. Eastern North Dakota to the St. Lawrence drainage, including northern Illinois, Indiana, Michigan, Indiana and Ohio.

34b Mouth oblique to almost horizontal and extending to or almost to the front of the eye. 35

35a Anal rays usually 9 to 11 36

35b Anal rays usually 7 or 8 44

36a Fins all heavily tipped with black pigment. (See Fig. 285, couplet 9a.) . . . . .
. . . . PRETTY SHINER,
Notropis bellus (Hay)
Young or slender individuals may key here.

36b Fins not all heavily tipped with black pigment 37

37a Front of dorsal fin above or slightly before or slightly behind front of pelvic fins. Fig. 318 38

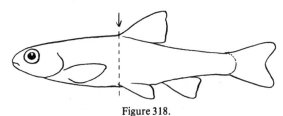

Figure 318.

37b Front of dorsal fin completely behind entire base of pelvic fins. Fig. 319 40

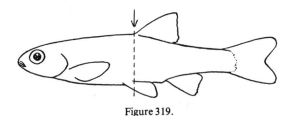

Figure 319.

38a Lateral band not developed on sides . . . .
. . . . COMMON SHINER,
Notropis cornutus (Mitchill)
Young individuals may key here. (See Fig. 282, couplet 9a.)
CRESCENT SHINER, *Notropis cerasinus* (Cope). Young individuals may key here. (See Fig. 283, couplet 9a.)
WHITE SHINER, *Notropis albeolus* Jordan. The adults are rather deep bodied but immature individuals may be more slender and will key here. (See Fig. 284, couplet 9a.)

38b Lateral band of pigment present on sides but may be faint or developed only on peduncle **39**

39a Lateral band of pigment more or less dusky but not black. Fig. 320. . . . . **SAFFRON SHINER,** *Notropis rubricroceus* (Cope)

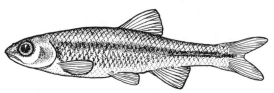

Figure 320.

Olivaceous with more or less red color; lateral band extends onto caudal fin; snout longer than eye diameter; maxillary reaches to or behind front of eye. Anal rays 8-9. Teeth 2,4-4,2. Length 4 inches. Headwaters of the Tennessee River.

SILVERBAND SHINER, *Notropis shumardi* (Girard). Fig. 321. Pale straw color with silvery band; snout shorter than diameter of eye; maxillary reaches to the front of the eye. Anal fin rays 8 or 9. Lateral line scales 34-37. Teeth 2,4-4,2. Length 2 1/2 inches. Iowa southward in Mississippi basin into eastern Oklahoma, Louisiana and Texas; present in Mobile Bay drainage.

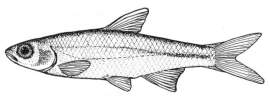

Figure 321.

WHITETAIL SHINER, *Notropis galacturus* (Cope). Fig. 322. Anal rays 8 or 9, individuals with 9 anal rays will key here. Females olivaceous, males steel blue above, white below. Caudal fin with distinct white area at base; dorsal fin with black blotch on posterior

Figure 322.

rays. Teeth 1,4-4,1. Length 5-6 inches. Ozark streams in Missouri and Arkansas and headwaters of Cumberland and Tennessee Rivers.

WARPAINT SHINER, *Notropis coccogenis* (Cope). Fig. 323. Anal rays 8-9, individuals with 9 anal rays will key here. Olivaceous above, silvery below. Similar to whitetail shiner but dark vertical bar just behind opercle. Spring males are rosy below and red on heads. Caudal fin with light basal area. Upper portion of dorsal fin black. Teeth 2,4-4,2. Length 5 inches. Headwaters of Cumberland and Tennessee Rivers.

Figure 323.

39b Lateral band very dark (black) and well developed. Fig. 324 . . . . **BLEEDING SHINER,** *Notropis zonatus* (Putnam)

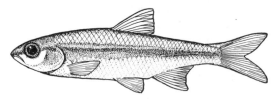

Figure 324.

Olivaceous with dorsal lateral stripes on back; no spots on fins. Maxillary does not quite reach front of eye. Males with bright red sides. Teeth 2,4-4,2. Length 5 inches. Ozark streams tributary to the lower Missouri River in Arkansas and Missouri, small tributaries of Mississippi River in eastern Missouri.

DUSKYSTRIPED SHINER, *Notropis pilsbryi* Fowler. Similar to the bleeding shiner but lateral band extends below lateral line; no dorsal lateral stripes present on back. Red pigmentation on body and fins less pronounced. Teeth 2,4-4,2. Length 4 inches. White River in Arkansas and Missouri, a few tributaries of the Arkansas and Red Rivers in Kansas, Arkansas and Oklahoma.

POPEYE SHINER, *Notropis ariommus* (Cope). (See couplet 27b.) Maxillary reaches front of eye. Usually slender, but stout individuals will key here.

TRICOLOR SHINER, *Notropis trichroistius* (Jordan and Gilbert). (See Fig. 291, couplet 13b.) Rather deep bodied, but more slender individuals will key here.

40a Scales before dorsal fin 29 to 30. Fig. 325 . . . . **REDFIN SHINER,** *Notropis umbratilis* (Girard)

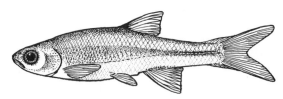

Figure 325.

Rather deep bodied, but immature individuals are more slender.

40b Scales before dorsal fin less than 29 . . . 41

41a Basal portion of caudal fin distinctly marked with white or cream color . . . . **SAILFIN SHINER,** *Notropis hypselopterus* (Gunther)
Blotch in dorsal fin; 2 rosy spots at base of caudal fin. Usually a deep bodied fish, but more slender individuals will key here. (See Fig. 279, couplet 8a.)

ALTAMAHA SHINER, *Notropis xaenurus* (Jordan). Dark steel blue above, silvery below; faint caudal spot continuous with lateral band; upper posterior part of dorsal fin black. Teeth 1,4-4,1. Length 3 inches. Altamaha River drainage, Georgia.

WHITEFIN SHINER, *Notropis niveus* Cope. Fig. 326. Dorsal fin with posterior blotch; median fins light tipped; caudal spot faint. Teeth 1,4-4,1. Coastal streams Virginia to South Carolina.

Figure 326.

41b Basal portion of caudal fin not marked with white or cream-colored area 42

42a Dorsal fin marked with a blotch . . . . **GREENFIN SHINER,** *Notropis chloristius* (Jordan and Brayton)
Similar to whitefin shiner but base of caudal fin not marked with white and no trace of a caudal spot. Fins pale at tips. Teeth 1,4-4,1. Length 3 inches. Santee River system, North and South Carolina.

42b Dorsal fin without any blotch 43

43a Scales before dorsal fin 20 or more. Fig. 327 . . . . **BLUNTNOSE SHINER,** *Notropis simus* (Cope)

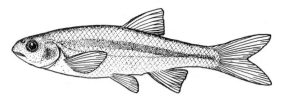

Figure 327.

Plain silvery. Teeth 1,4-4,1. Length 3 1/2 inches. Rio Grande River drainage, Texas and New Mexico.

SHARPNOSE SHINER, *Notropis oxyrhynchus* Hubbs and Bonham. Fig. 328. Pale silvery sides with faint lateral band posteriorly. Teeth 2,4-4,2. Length 2 1/2 inches. Central Texas.

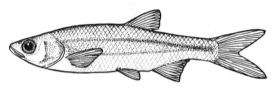

Figure 328.

43b Scales before dorsal fin less than 20. Fig. 329. . . . . SANDBAR SHINER, *Notropis scepticus* (Jordan and Gilbert)

Figure 329.

Greenish silvery with silvery lateral band. Teeth 2,4-4,1. Length 3 inches. North and South Carolina to Florida.

DUSKY SHINER, *Notropis cummingsae* Myers. Fig. 330. Silvery with dark lateral band. Teeth 2,4-4,1. Length 3 inches. North and South Carolina.

Figure 330.

HIGHFIN SHINER, *Notropis altipinnis* (Cope). Immature individuals with more slender bodies may key here. (See Fig. 278, couplet 8a.)

EMERALD SHINER, *Notropis atherinoides-* Rafinesque. (See Fig. 313, couplet 29a.) Heavy bodied individuals will key here.

FLAGFIN SHINER, *Notropis signipinnis* Bailey and Suttkus. (See Fig. 280, couplet 8a.) Silvery with very broad lateral band. Slender individuals will key here.

RIO GRANDE SHINER, *Notropis jemezanus* (Cope). Fig. 331. Pale silvery with a lateral band best developed posteriorly. Lateral line well below lateral band. Teeth, 2,4-4,2. Length 3 inches. Rio Grande River drainage and southward.

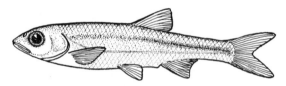

Figure 331.

44a Lateral line scales 40 or more than 40 . . 45

44b Lateral line scales less than 40 47

45a Basal portion of caudal fin not milky white. Fig. 332. . . . . ALABAMA SHINER, *Notropis callistius* (Jordan)

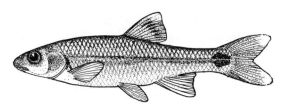

Figure 332.

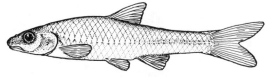

Figure 333.

Olivaceous; tip of dorsal fin white; caudal fin red but milky at tips; very large caudal spot. Similar to blacktail shiner, but has more scales in lateral line. Teeth 1,4-4,1. Length 4 inches. Mobile Bay drainage Alabama and Georgia.

45b Basal portion of caudal fin milky white 46

46a Snout slightly longer than diameter of eye; dark bar on sides just behind opercle. (See Fig. 323, couplet 39a.) . . . . WARPAINT SHINER, *Notropis coccogenis* **(Cope)**

Anal rays 8-9; individuals with 8 anal rays will key here.

46b Snout almost twice the diameter of the eye; no dark bar on sides back of opercle. (See Fig. 322, couplet 39a.) . . . . WHITETAIL SHINER, *Notropis galacturus* **(Cope)**

Anal rays 8-9; individuals with 8 anal rays will key here.

47a No dark spot on or at the base of the caudal fin 48

47b A more or less distinct dark spot on or at the base of the caudal fin (may be fused with lateral band). 54

48a Upper lip considerably below level of lower margin of eye; mouth almost horizontal. Fig. 333. . . . . SABINE SHINER, *Notropis sabinae* **Jordan and Gilbert**

Pale olive with poorly developed lateral band; mouth nearly horizontal and included. Maxillary extends past anterior margin of pupil. Anal rays 7. Scales in lateral line 34-37. Teeth 4-4 or 1,4-4,1. Black River, southeastern Missouri, northeastern Arkansas, Neches River, Texas; Sabine River, Texas and Louisiana, Little Calcasieu River, Louisiana.

BLUNTNOSE SHINER, *Notropis simus* (Cope). (See Fig. 327, couplet 43a.) Individuals with 8 anal rays will key here.

PROSPERINE SHINER, *Notropis prosperpinus* (Girard). Fig. 334. Silvery with dark lateral band, black streak from chin to isthmus; maxillary extends almost to front of orbit; anal rays (7) 8. Teeth 4-4. Rio Grande River drainage, Texas.

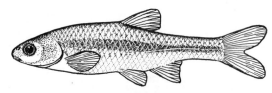

Figure 334.

PLATEAU SHINER, *Notropis lepidus* (Girard). Similar to prosperine shiner black streak under chin extends only to below eye. Teeth 1,4-4,1. Texas.

48b Upper lip on or above level of lower margin of eye; mouth or less oblique . . 49

49a Maxillary extending past front of eye; lateral band very dark. Fig. 335. . . . . BIGEYE SHINER, *Notropis boops* **Gilbert**

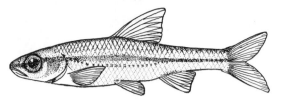

Figure 335.

Slim olivaceous minnow with dusky sides. Lateral band passes through eye and over snout. Scales large, 35-36 in lateral line. Teeth 1,4-4,1. Length 2 1/2 inches. Lower Ohio River to the south in Tennessee, south and west in the Mississippi River basin to Kansas, Missouri, Arkansas and Oklahoma.

SAFFRON SHINER, *Notropis rubricroceus* (Cope). (See Fig. 320, couplet 39a.) Individuals with 8 anal rays will key here.

49b   Maxillary not extending back of front margin of eye.................50

50a   Lateral band dark and may be narrow. Fig. 336........................ .... NEW RIVER SHINER, *Notropis scabriceps* (Cope)

Figure 336.

Olivaceous with dusky lateral band which extends forward onto snout. Teeth 2,4-4,2. Length 3 inches. Upper drainage of the Kanawha River, Virginia and West Virginia.

KIAMICHI SHINER, *Notropis ortenburgeri* Hubbs. (See Fig. 310, couplet 27a.) Deep bodied individuals will key here.

BLACKNOSE SHINER, *Notropis heterolepis* Eigenmann and Eigenmann. (See Fig. 316, couplet 33b.) Lateral line variable; individuals with complete lateral line will key here.

50b   Lateral stripe rather pale and mostly posterior 51

51a   Anal fin rays usually 7. Fig. 337....... ....SAND SHINER, *Notropis stramineus* (Cope)

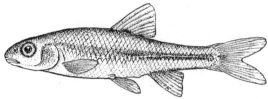

Figure 337.

Silvery with thin mid-dorsal streak and black dash in center of dorsal base. Pigment may be present at base of anal fin. Teeth 4-4. Length 2 1/2 inches. Southeastern Saskatchewan, southern Manitoba, Montana, North Dakota east through Great Lakes to the St. Lawrence, south through Colorado, Oklahoma to Texas and Mexico; northern two-thirds of Missouri east to Tennessee River in Alabama and northward through the Mississippi drainage.

SKYGAZER SHINER, *Notropis uranoscopus* Suttkus. Resembles sand shiner but has a small wedge-shaped caudal spot which will key it in couplet 57a. Tallapoosa, Cahaba and Alabama river systems in Alabama.

RIVER SHINER, *Notropis blennius* (Girard). Fig. 338. Pale olive, no pigment at base of anal fin; dorsal streak prominent. Eye small, much less than snout length. Scales in the lateral line usually 35 or more. Teeth (2)1,4-4,1(2). Southern Alberta, Saskatchewan and southwestern Manitoba, eastern Wyoming to Pennsylvania, south to Tennessee, Oklahoma, Arkansas, Louisiana and Texas.

Figure 338.

FLUVIAL SHINER, *Notropis edwardraneyi* Suttkus and Clemmer. Fig. 339. Silvery in color, similar to the river shiner. Eye larger, equal to or greater than snout length. Scales in lateral line usually 34 or less. Teeth 2,4-4,2. Alabama and Tombigee Rivers, Alabama.

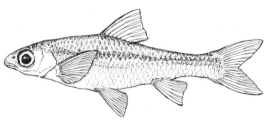

Figure 339.

51b   Anal rays usually 8. . . . . **52**

52a   Scales (or scale rows) before dorsal fin 16 or more. Fig. 340. . . . . **BIGMOUTH SHINER,** *Notropis dorsalis* (Agassiz)

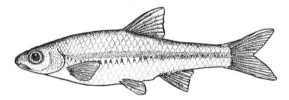

Figure 340.

Olivaceous above and silvery below; mid-dorsal streak thin. Nape only partially scaled or naked in some populations. Snout V-shaped when viewed from above. Teeth 1,4-4,1. Length 2 1/2 inches. North Dakota, Colorado east to the St. Lawrence and Ohio River drainages and south to Mexico.

52b   Scales before dorsal fin less than 16 . . . **53**

53a   Mid-dorsal streak present although may be thin and pale. . . . . **MIMIC SHINER,** *Notropis volucellus* (Cope)

Heavy bodied individuals may key here. (See Fig. 296, couplet 20a.)

53b   No mid-dorsal streak. Fig. 341 . . . . **GHOST SHINER,** *Notropis buchanani* Meek

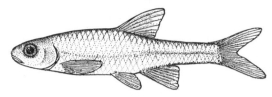

Figure 341.

Very similar to mimic shiner but more pale and with higher lateral line scales. Pigmented at base of anal fin. Teeth 4-4. Length 2 1/2 inches. Iowa through the Ohio River drainage south to Alabama and west to Texas and Mexico.

PALLID SHINER, *Notropis amnis* Hubbs and Greene. Fig. 342. Silvery in color, with the aspect of *Hybopsis,* but lacks the barbel. Snout overhangs small horizontal mouth. Scales in lateral line 36-38. Teeth 1,4-4,1. Length 2 1/2 inches. Lowlands of the Mississippi River from Minnesota and Wisconsin south to Oklahoma, Louisiana and into eastern Texas; also present in the Ohio River drainage, Indiana.

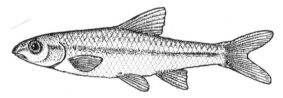

Figure 342.

54a   Anal rays usually 7 (except in the blacktail shiner with a larger black caudal spot, which has 7 to 8 anal rays) **55**

54b   Anal rays usually 8. . . . . **61**

55a Sides with scattered black spots above lateral line. Fig. 343. . . . . **CHIHUAHUA SHINER,** *Notropis chihuahua* Woolman

Figure 343.

Light brown with spots. Teeth 4-4. Length 2 inches. Mexico north into the Rio Grande drainage of Texas.

55b Sides above lateral line not marked with scattered black spots 56

56a Caudal spot larger than eye . . . . **BLACKTAIL SHINER,** *Notropis venustus* (Girard)
Rather deep bodied, but slender individuals with only 7 anal rays will key here. (See couplet 5a.)

56b Caudal spot smaller than eye 57

57a Snout shorter than diameter of eye. Fig. 344. . . . . **BLACKSPOT SHINER,** *Notropis atrocaudalis* Evermann

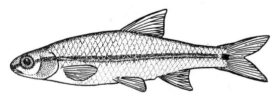

Figure 344.

Olivaceous on back with white ventral surface. Prominent black lateral band extending through eye and around snout, ending in well-defined caudal spot. Eye large, snout blunt.

Teeth 4-4. Length 2 1/2 inches. Limited range; eastern Texas, western Louisiana, extreme southeastern Oklahoma and southwestern Arkansas.

COOSA SHINER, *Notropis xaenocephalus* (Jordan). (See Fig. 279, couplet 20b.) Deeper bodied individuals will key here. Resembles the weed shiner but has 13 scales before the dorsal fin instead of 15 and has more decurved lateral line.

WEED SHINER, *Notropis texanus* (Girard). (See Fig. 347, couplet 59a.) Individuals with short snouts will key here.

SKYGAZER SHINER, *Notropis uranoscopus* Suttkus. May key here as it has a small wedge-shaped caudal spot. (See couplet 51a.)

57b Snout equal to or longer than diameter of eye. 58

58a Lateral band does not pass through eye and over snout. Fig. 345 . . . . **RED RIVER SHINER,** *Notropis bairdi* Hubbs and Ortenburger

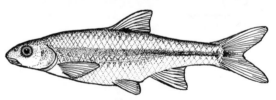

Figure 345.

Pale silver, grayish above; pigment outlines dorsolateral scales. Anal rays 7. Teeth 4-4. Length 2 1/2 inches. Red River in Oklahoma and Texas.

SMALLEYE SHINER, *Notropis buccala* Cross. Similar to Red River shiner except the snout is longer and the mouth smaller. Teeth 4-4. Length 2 inches. Brazos River, Texas.

CHUB SHINER, *Notropis potteri* Hubbs and Bonham. Fig. 346. Olivaceous, dusky above and silvery below; iris red in living specimens. Eye small, usually slightly more

Figure 346.

than 4 times in head length. Anal fin rays 7. Teeth (1)2,4-4,2(1). Length 2 1/2 inches. Colorado and Brazos Rivers in Texas, Red River in Texas, Oklahoma and Louisiana; lower Mississippi River drainage in Louisiana.

HIGHSCALE SHINER, *Notropis hypsilepis* Suttkus and Raney. Deeper bodied individuals will key here. See couplet 20b.

58b **Lateral band passes through eye and on snout** **59**

59a **Lateral band touches chin. Fig. 347** .....
.... **WEED SHINER,**
Notropis texanus **(Girard)**

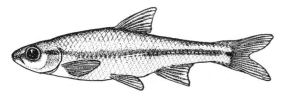

Figure 347.

Olivaceous above, silvery below; with a broad dark lateral band and small dark caudal spot. Anal fin rays 7. Teeth 2,4-4,2. Length 2 1/2 inches. Headwaters of the Red River, Minnesota to Lake Michigan in the Great Lakes basin, southward through eastern Iowa and Wisconsin in the Mississippi River drainage to central Texas; the Gulf coast drainage from Texas to the Apalachicola drainage in Georgia and Florida.

59b **Lateral band confined to snout and does not touch chin** **60**

60a **Snout longer than diameter of eye and rather pointed, protruding slightly beyond upper lip. Fig. 348** ....
.... **COASTAL SHINER,**
Notropis petersoni **Fowler**

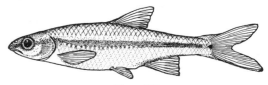

Figure 348.

Dusky above, silvery below. Teeth 2,4-4,2. Length 2 inches. North Carolina into Alabama, central and western Florida.

ROUGH SHINER, *Notropis baileyi* Suttkus and Raney. Fig. 349. Dusky above and silvery below with a lateral band from snout to prominent spot at base of caudal fin. Teeth 2,4-4,2. Length 2 1/2 inches. Alabama River, Tombigbee River, Escambia River and Pascagoula River drainages of Mississippi and Alabama; Bear Creek, Tennessee River, Alabama.

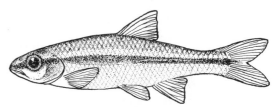

Figure 349.

BLACKSPOT SHINER, *Notropis atrocaudalis* Evermann. (See Fig. 344, couplet 57a.) Individuals with a longer snout will key here.

BURRHEAD SHINER, *Notropis asperifrons* Suttkus and Bailey. (See Fig. 303, couplet 21a.) Deeper bodied individuals will key here.

60b **Snout short, not longer than diameter of eye and rather rounded, not extending beyond upper lip. Fig. 350** ....
.... **WHITEMOUTH SHINER,**
Notropis alborus **Hubbs and Raney**

Figure 350.

Dusky above and silvery below. Teeth 4-4. Length 2 inches. Roanoke River in Virginia to Santee River in South Carolina.

SWALLOWTAIL SHINER, *Notropis procne* (Cope). (See Fig. 297, couplet 20a.) Rather heavy bodied individuals will key here. Snout is short but rather sharp instead of rounded.

TOPEKA SHINER, *Notropis topeka* Gilbert. (See Fig. 276, couplet 7b.) Eye smaller than in *N. alborus*. Slender individuals will key here.

RAINBOW SHINER, *Notropis chrosomus* (Jordan). (See Fig. 353, couplet 63a.) Individuals with 7 anal rays will key here.

61a Mouth very oblique and small, extending only about 2/3 the length of the snout. Fig. 351. . . . .
. . . . IRONCOLOR SHINER, *Notropis chalybaeus (Cope)*

Figure 351.

Dark above and silvery below with a very black lateral band which extends through eye and snout. Small detached black spot at base of caudal fin. Teeth 2,4-4,2. Length 2 inches. Coastal lowlands, New Jersey to eastern Texas and north in Mississippi drainage to Iowa.

CAPE FEAR SHINER, *Notropis mekistocholas* Snelson. Fig. 352. Similar to the

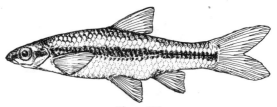

Figure 352.

whitemouth and swallowtail shiner but with 8 rather than 7 anal fin rays. Differs from all other members of the genus *Notropis* in having an elongate intestine, not S-shaped. Olivaceous with white ventral surface, prominent black lateral band; wedge-shaped caudal spot detached or weakly connected to lateral band. Peritoneum black. Teeth 4-4. Length 2 inches. Cape Fear River drainage, North Carolina.

61b Mouth moderate length, reaching to or almost to the front of eye. 62

62a Spot at base of caudal fin very dark . . . 63

62b Spot at base of caudal fin rather pale, not large 65

63a Caudal spot not detached from lateral band. Fig. 353. . . . .
. . . . RAINBOW SHINER, *Notropis chrosomus* (Jordan)

Figure 353.

Dusky above, silvery below. Males with scarlet band above lateral band and scarlet bar across anal, dorsal and caudal fins. Teeth 2,4-4,2. Length 3 inches. Alabama, Coosa and Cahaba rivers, Mobile Bay drainage.

COOSA SHINER, Notropis xaeno-

cephalus (Jordan). Individuals with 8 anal rays will key here. (See Fig. 299, couplet 20b.)

63b Caudal spot detached from lateral band 64

64a Caudal spot triangular; scales at base of anal fin pigmented. Fig. 354 . . . . **WEDGESPOT SHINER,** *Notropis greenei* **(Hubbs and Ortenburger)**

Figure 354.

Dusky above, pale below. Anal fin rays 8. Scales in the lateral line 35-38. Teeth 2,4-4,2. Length 2 1/2 inches. Ozark drainage of Missouri, Arkansas and eastern Oklahoma.

64b Caudal spot not triangular but more or less rounded; scales at base of anal fin not pigmented. Fig. 355 . . . . **SPOTTAIL SHINER,** *Notropis hudsonius* **(Clinton)**

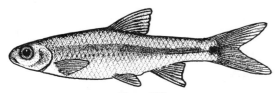

Figure 355.

Pale silvery with more or less dusky lateral band. Teeth, shape and markings variable in different regions. Caudal spot may be faint or absent in sub-species *N. h. amarus* (Girard) in Delaware and Potomac River drainages and much variation occurs in the sub-species *N. h. saludanus* (Jordan and Brayton) in the coastwide streams southward to Georgia. Distribution from Mackenzie River, Northwest Territories, Alberta, North Dakota to the Hud-

son River and south to Iowa, Illinois and Atlantic coastal streams to Georgia. Length 5 inches.

65a Front of dorsal fin distinctly before front of pelvic fins. Fig. 356 . . . . **ARKANSAS RIVER SHINER,** *Notropis girardi* **Hubbs and Ortenburger**

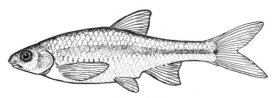

Figure 356.

Dusky above, pale below; mouth suterminal and almost horizontal. Anal fin rays 8. Scales in the lateral line 33-37. Teeth 4-4. Length 2 inches. Arkansas River drainage, Kansas, Arkansas and Oklahoma.

65b Front of dorsal fin over or behind front of pelvic fins. Fig. 357 . . . . **TEXAS SHINER,** *Notropis amabilis* **(Girard)**

Figure 357.

Olivaceous above, silvery below; front of dorsal fin above or slightly behind front of pelvic fins. Teeth 2,4-4,2. Length 2 1/2 inches. Central Texas southward into Mexico.

SILVERBAND SHINER, *Notropis shumardi* (Girard). Fig. 358. Individuals with 8 anal rays may key here. (See Fig. 321, couplet 39a.)

YELLOWFIN SHINER, *Notropis lutipinnis* (Jordan and Brayton). Fig. 359. Dusky green above, silvery below; base of dorsal fin behind base of pelvic fins. Teeth 2,4-4,2.

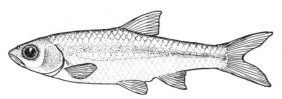

Figure 358.

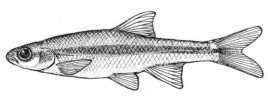

Figure 359.

Length 3 inches. Chattahoochee River, Alabama and Georgia.

BLUE SHINER, *Notropis caeruleus* (Jordan). Fig. 360. Bluish above, silvery white

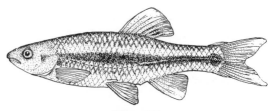

Figure 360.

Length 2 1/2 inches. Santee River drainage, North and South Carolina to Savannah River.

OCMULGEE SHINER, *Notropis callisema* (Jordan). Bluish above, silvery below; large dark spot on last rays of dorsal fin; white tips on dorsal, anal, and caudal fins; dark lateral band ending in caudal spot. Teeth 4-4. Length 2 3/4 inches. Altamaha River drainage, Georgia.

BLUESTRIPE SHINER, *Notropis callitaenia* Bailey and Gibbs. Very similar to *N. callisema*. Caudal spot confluent with lateral band as in blacktail shiner. Teeth 1,4-4,1.

below; lateral band continuous with inconspicuous caudal spot; faint blotch in upper part of dorsal fin; milky tips on dorsal, anal and caudal fins. Snout rather pointed and overhanging. Teeth 1,4-4,1. Length 3 inches, Coosa and Cahaba Rivers in Mobile Bay drainage, Alabama and Georgia.

OHOOPEE SHINER, *Notropis leedsi* Fowler. Slender individuals will key here. (See Fig. 277, couplet 7b.)

BLACKTAIL SHINER, *Notropis venustus* (Girard). Slender forms with faint caudal spots will key here. (See Fig. 273, couplet 5a and couplet 56a.)

SUCKER FAMILY
Catostomidae

This family is closely allied to the minnow family. The suckers are softrayed fishes and possess a toothless and more or less sucker-like protractile mouth with thick lips. Lips may be covered with *plicae* which are longitudinal folds separated by fine creases or may be covered by numerous papillae. The last pharyngeal arch bears a row of numerous comb-like teeth (See Fig. 384.) which distinguishes the suckers from the minnows which have either two rows of

teeth, or one row with only a few (5) teeth. Suckers usually have more than 10 dorsal rays, whereas, most native minnows have no more than 10 dorsal rays.

The heads are naked, and the body is covered with smooth cycloid scales. They are extremely bony as the ribs, including a set of accessory ribs, are distributed from the head to the tip of the tail. Otherwise, the flesh of most members of this family is quite edible.

This family contains many species which are quite widely distributed in the United States and furnish an important group of our so-called forage fishes. They are mostly omnivorous, feeding on the bottom where they eat a large variety of animal matter as well as some plant material.

They spawn in the spring, many species making spectacular spawning "runs" up small tributary streams. They are very prolific, and the eggs are scattered at random and develop without any parental care. During spawning season the males in some species may develop tubercles on the head, brighter colors and elongated anal fins.

1a Dorsal fin long with 25 to 40 rays. 2

1b Dorsal fin short with 10 to 18 rays 4

2a Lateral line scales more than 50; eyes in back part of head; head small and abruptly more slender than body; body is 6-7 times length of head . . . . *Cycleptus*
BLUE SUCKER, *Cycleptus elongatus* (Lesueur). Fig. 361. Dark back and dusky silvery on sides. Reaches a length of over 2 feet. In large rivers from southern Minnesota and Wisconsin to Tennessee, Mobile Bay drainage, and Mexico.

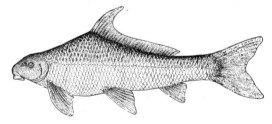

Figure 361.

2b Lateral line scales less than 45; eyes in front part of head; head large and not abruptly more slender than body 3

3a Distance from eye to lower posterior angle of preopercle about 3/4 distance to upper corner of gill cleft; sub-opercle widest at middle; pharyngeal arch thick, triangular in cross section . . . .
. . . . BUFFALOFISHES, *Ictiobus*
This genus contains three species. They are large golden or reddish brown fishes with deep bodies.

LARGEMOUTH BUFFALOFISH, *Ictiobus cyprinellus* (Valenciennes). Fig. 362. Mouth large with upper lip about level with lower margin of eye. Reaches a length of about 3 feet. Saskatchewan and North Dakota to Lake Erie and south to Alabama and northern Texas. Introduced into reservoirs of Gila River, Arizona and Los Angeles Aqueduct system, southern California.

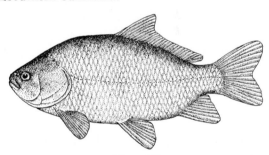

Figure 362.

SMALLMOUTH BUFFALOFISH, *Ictiobus bubalus* (Rafinesque). Mouth small (Fig. 363) with upper lip far below lower margin of eye; back quite elevated. Reaches a length of 2 1/2 feet. Southern Minnesota to Michigan and south to Mexico.

Figure 363.

BLACK BUFFALOFISH, *Ictiobus niger* (Rafinesque). Mouth same as for smallmouth buffalofish, but back not much elevated. Reaches length of 3 feet. Southern Minnesota and Michigan to Texas.

3b **Distance from eye to lower posterior angle of preopercle about equal to distance to upper corner of gill cleft; subopercle widest below middle (Fig. 364); pharyngeal arch almost paper thin......** .... **CARPSUCKERS,** *Carpiodes*

Figure 364.

Color more or less silvery. Some may reach a length of over 20 inches. This genus contains three species in the United States.

QUILLBACK CARPSUCKER, *Carpiodes cyprinus* (Lesueur). Fig. 365. No nipple-like structure on lower lip (Fig. 366);

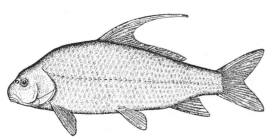

Figure 365.

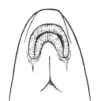

Figure 366.

anterior dorsal rays elongated, or as long as base of fin; lateral line scales 37-40; nostril posterior to middle of mouth (Fig. 364).

HIGHFIN CARPSUCKER, *Carpiodes velifer* (Rafinesque). Nipple-like structure on lower lip (Fig. 367); anterior dorsal rays much elongated; mouth mostly posterior to nostrils (Fig. 368); front of upper lip under nostrils; lateral line scales 33-37. Minnesota to Pennsylvania and south to Tennessee.

Figure 367.

Figure 368.

CARPSUCKER, *Carpiodes carpio* (Rafinesque). Fig. 370. Nipple-like structure on lower lip (Fig. 367); anterior dorsal rays only slightly elongated, not reaching more than 1/2 the base of the fin; mouth mostly posterior to nostrils; front of upper lip almost under nostril (Fig. 369); lateral line scales 33-37. Montana to Pennsylvania and south into Mexico and northern Florida.

Figure 369.

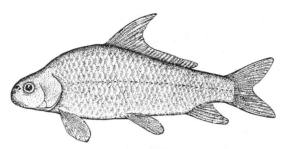

Figure 370.

4a Lateral line absent or incomplete in adults **5**

4b Lateral line complete in adults **6**

5a Lateral line partly or almost complete; each scale with a distinct spot, forming a pattern of rows of dotted lines on sides; mouth inferior and horizontal . . . . *Minytrema*
SPOTTED SUCKER, *Minytrema melanops* (Rafinesque). Fig. 371. Silvery, distinctly characterized by a spot on each scale. Reaches a length of about 18 inches. Larger streams, southern Minnesota to Pennsylvania and south to eastern Texas and Florida.

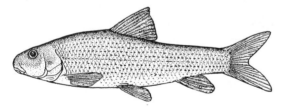

Figure 371.

5b Lateral line absent; sides of young with longitudinal stripe which breaks up into blotches in adults; mouth sub-terminal and oblique . . . . **CHUBSUCKERS,** *Erimyzon*
Small silvery fishes reaching a length of about 10 inches.
LAKE CHUBSUCKER, *Erimyzon sucetta* (Lacépède). Fig. 372. Longitudinal scale

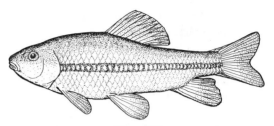

Figure 372.

rows 36-38. Southern Wisconsin to New England and south to Florida and eastern Texas.
CREEK CHUBSUCKER, *Erimyzon oblongus* (Mitchill). Longitudinal scale rows 39-41. Wisconsin to New England and south to Alabama and Texas.
SHARPFIN CHUBSUCKER, *Erimyzon tenuis* (Agassiz). Fig. 373. Differs in that first dorsal ray is as long as base of dorsal fin. Dark lateral band present in young individuals and continues around snout. Gulf drainage of Mississippi, Alabama, and western Florida.

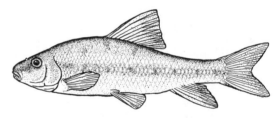

Figure 373.

6a Pronounced hump on back just behind head . . . . *Xyrauchen*
HUMPBACK SUCKER, *Xyrauchen texanus* (Abbott). Fig. 374. Rather olivaceous. Reaches length of 2 feet. Colorado River, Wyoming southward.

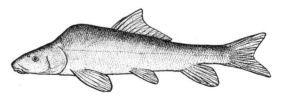

Figure 374.

6b No pronounced hump on back just behind head. 7

7a Scales less than 55 in lateral line and not crowded anteriorly 8

7b Scales more than 55 in lateral line and crowded anteriorly. 12

8a Top of head between eyes concave; swim bladder in 2 parts. . . . . **HOGSUCKERS,** *Hypentelium*
Dusky silver, mottled with black. Reaches a length of about 10-12 inches.
      NORTHERN HOGSUCKER, *Hypentelium nigricans* (Lesueur). Fig. 375. Dorsal rays 11; lateral line scales 46; total pectoral rays for both sides 34; dark saddle crosses back before dorsal fin. Minnesota to New York and south to Oklahoma and Alabama.

Figure 375.

ROANOKE HOGSUCKER, *Hypentelium roanokense* Raney and Lachner. Dorsal rays 11, but differs from northern hogsucker in having 41 lateral line scales and in having total pectoral rays for both sides 31, dark saddle not developed before dorsal fin. Headwaters of Roanoke River, Virginia.
      ALABAMA HOGSUCKER, *Hypentelium etowanum* (Jordan). Fig. 376. Differs in having dorsal rays 10 (9-11). Dorsal fin with black tip anteriorly. Chattahoochee River system of Alabama and Georgia, and throughout the Mobile Bay drainage.

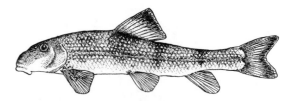

Figure 376.

8b Top of head usually convex between eyes; air bladder obsolete or in 3 parts 9

9a Premaxillary not protractile; lower lip separated from upper lip by lateral notches and divided into two separate lobes (Fig. 377) . . . . *Lagochila*

Figure 377.

HARELIP SUCKER, *Lagochila lacera* Jordan and Brayton. Length up to 18 inches. Rare in larger streams of central Mississippi valley.

9b Premaxillary protractile; lower lip not divided from upper lip by lateral notches and not completely divided into two lobes . . . . (*Moxostoma,* **REDHORSES**) 10

10a Lower lip with plicae (longitudinal ridges) which are broken posteriorly by diagonal creases which may form oval or round papillae (Figs. 378, 379). . . . . **TORRENT SUCKER,** *Moxostoma (Thoburnia) rhotheca* (Thoburn)

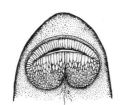

Figure 378.

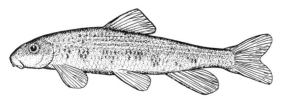

Figure 379.

Somewhat mottled with lateral streaks on sides; halves of lower lip slightly rounded posteriorly and separated by deep notch (Fig. 378). Headwaters of the James, Kanawha, and perhaps the Potomac Rivers, Virginia.

RUSTYSIDE SUCKER, *Moxostoma (Thoburnia) hamiltoni* (Raney and Lachner). Similar to torrent sucker; no pigment in dorsal fin; halves of lower lip separated by deep notch but rounded posteriorly. Roanoke River system, Virginia.

BLACKFIN SUCKER, *Moxostoma atripinne* Bailey. Fig. 380. Black blotch on distal half of first 5-6 dorsal fin rays; lower lip with posterior margin almost straight and with a shallow marginal groove; plicae of lips broken by a few transverse creases and with no papillae. Barren River system, Greene River, Tennessee.

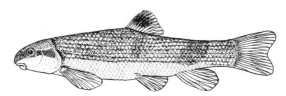

Figure 380.

10b Lower lip with plicae which may be smooth or may be broken by transverse creases. 11

11a Scales around peduncle 12, usually 5 above and 5 below lateral line; more or less reddish brown and silvery, fins tend to be reddish. Length of 24 inches or more. Fig. 381. . . . .
. . . . SILVER REDHORSE, *Moxostoma anisurum* (Rafinesque)

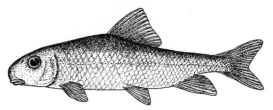

Figure 381.

Halves of lower lip thin and meet at an acute angle (Fig. 382A); plicae broken by transverse creases; dorsal fin rounded at tip. Saskatchewan to St. Lawrence drainage and south to Missouri and Alabama.

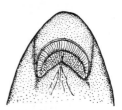

Figure 382A.

SHORTHEAD REDHORSE, *Moxostoma macrolepidotum* (Lesueur). Fig. 383. Halves of lower lip meet in almost a straight line (Fig. 382B); plicae of lower lips are almost smooth but may be broken in corners. Dorsal fin pointed, about 13 rays. Pharyngeal arch thin, teeth fine and comb-like (Fig. 384A).

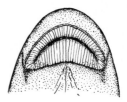

Figure 382B.

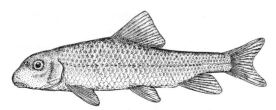

Figure 383.

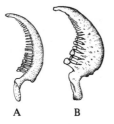

A   B

Figure 384.

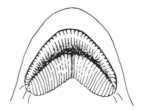

Figure 385.

nesota to Ontario and south to Alabama and Oklahoma in the Mississippi drainage and from the Great Lakes drainage to western Florida.

GOLDEN REDHORSE, *Moxostoma erythrurum* (Rafinesque). Plicae of lips with no transverse creases; lower halves meet in almost straight line (Fig. 386). Similar to greater

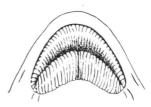

Figure 386.

redhorse except scale bases have no black spots and the dorsal fin is scarcely falcate and is black on sharp tip and near the margin; also differs in scales on peduncle. Minnesota and southern Ontario south to Oklahoma and Georgia.

NEUSE REDHORSE, *Moxostoma lachrymale* (Cope). Resembles the golden redhorse so closely that it may be a sub-species. Neuse River, North Carolina.

BLACKTAIL REDHORSE, *Moxostoma poecilurum* Jordan. Fig. 387. Similar to the northern redhorse except the lower lobe of the caudal fin is black and is narrower and longer

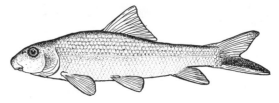

Figure 387.

Wide spread, Mackenzie River system to James Bay and south to Colorado, Oklahoma and Chesapeake Bay.

CAROLINA REDHORSE, *Moxostoma coregonus* (Cope). Similar to the shorthead redhorse but with a projecting snout and a deep median crease in lower lip. Catawba and Yadkin River systems, North Carolina.

RIVER REDHORSE, *Moxostoma carinatum* (Cope). Body large and thick; dorsal fin somewhat falcate; upper lobe of caudal fin is much longer than lower lobe. Differs from most redhorses in very heavy pharyngeal arch, triangular in cross section and in large molariform pharyngeal teeth (Fig. 384B). Minnesota to St. Lawrence River and south to Kansas, Mississippi, Alabama and western Florida.

SUCKERMOUTH REDHORSE, *Moxostoma pappillosum* (Cope). More or less silvery; lower lip with a deep median cleft and quite papillose. Coastal streams from Roanoke River, Virginia to Georgia.

BLACK REDHORSE, *Moxostoma duquesnei* (Lesueur). Rather slender; scales in lateral line 44-47 instead of 39-45 as in most other redhorses; 10 rays in pelvic fin instead of the usual 9 pelvic rays. Plicae of lips smooth or only slightly broken (Fig. 385). Southern Min-

than the upper lobe. Coastal streams of the Gulf of Mexico from eastern Texas to Florida.

**11b Scales around peduncle 16, unusually 7 above and 7 below lateral line when counted from side ....
...... GRAY REDHORSE, *Moxostoma congestum* (Baird and Girard)**

Resembles northern redhorse but differs in number of peduncle scales and has a rather low dorsal fin with 11-12 rays instead of 13. Halves of lower lip meet at acute angle with a deep median groove; lips plicate but broken into papillae in corners. Anal and distal half of dorsal fins dusky. Central Texas to Rio Grande River.

BLACK JUMPROCK, *Moxostoma cervinum* (Cope). Fig. 388. Dark saddles on back and light longitudinal streaks on sides; dorsal and caudal fins with black tips. Posterior margin of lower lip concave. Coastal streams from James to Neuse Rivers, Virginia.

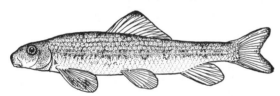

Figure 388.

STRIPED JUMPROCK, *Moxostoma rupiscartes* Jordan and Jenkins. Very similar to black jumprock except tips of fins are dusky instead of black; may retain juvenile blotches on sides and back. North Carolina southward in Santee, Savannah, Altamaha and Chattahoochee River systems.

GREATER JUMPROCK, *Moxostoma lachneri* Robbins and Raney. Related to striped jumprock with about 8 lateral streaks but has 12 dorsal rays instead of 10-11. Lower lobe of caudal fin dusky or black except for milky white lower ray as in blactail redhorse. Apalachicola River system, Georgia.

SMALLFIN REDHORSE, *Moxostoma robustum* (Cope). Lateral streaks as in striped

jumprock but head is shorter and lacks black pigment on tip of dorsal fin. Lower lip concave posteriorly, plicae may be slightly broken. Yadkin to Altamaha River systems in North Carolina to Georgia.

BIGEYE JUMPROCK, *Moxostoma ariommum* Robbins and Raney. Body with lateral streaks; lower lip very papillose. Differs from other redhorses as very concave between eyes. Upper Roanoke River, Virginia.

COPPER REDHORSE, *Moxostoma hubbsi* Legendre. Resembles river redhorse except body is heavier and shorter. Lake Ontario and St. Lawrence River drainages.

GREATER REDHORSE, *Moxostoma valenciennesi* Jordan. Lips with plicae not broken by transverse creases, lower halves meet at an angle. Resembles golden redhorse but dorsal fin is not falcate and has a whitish tip; scale bases lack dark spots. Minnesota, and Great Lakes and St. Lawrence drainages south to Iowa and Illinois but absent in most of the Ohio River drainage.

**12a Mouth terminal, lips rather thin with smooth or weakly broken plicae; lower lip divided by very wide median notch (Fig. 390) ....
.... *Chasmistes***

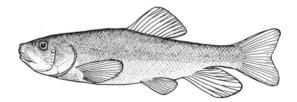

Figure 389.

A small group of western suckers with rather humped snouts; no axillary process present at the base of the pelvic fins. Rather restricted in their distribution.

SHORTNOSE SUCKER, *Chasmistes brevirostris* Cope. Fig. 389. Dusky above, pale below. Lateral line scales 70-80. Length up 18 inches. Klamath Lake, Oregon.

JUNE SUCKER, *Chasmistes liorus* Jordan. Scales on back with small punctulations; lateral line scales 58-65. Lower lip with broken plicae and with wide median notch (Fig. 390). Length 18 inches. Utah Lake, Utah.

Figure 390.

CUI-UI, *Chasmistes cujus* Cope. Pale olive above, very similar to June sucker except lateral line has 13-14 scales above instead of 7-11 scales. Restricted to Pyramid Lake, Nevada.

12b Mouth usually subterminal; lips rather thick, lower lip usually strongly papillose. SUCKERS, *Catostomus* 13
This genus contains a large number of species usually with dark backs shading to silvery below. Some species develop reddish streaks on sides during spawning season. Several former western genera have been recently included in this genus which now includes the mountain suckers, *Pantosteus,* and the lost sucker, *Deltistes.* Only several species are widespread east of the Rockies but many species have developed in the isolated stream systems west of the Rockies. Many suckers reach a length of 18 inches or more.

**13a Lower lip separated from upper lip by distinct lateral notches Figs. 391A, B, C
. . . . (*Pantosteus*) *Catostomus***

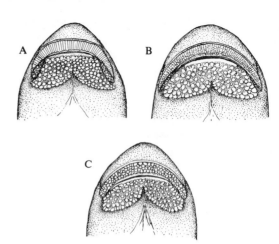

Figure 391.

MOUNTAIN SUCKER, *Catostomus (Pantosteus) platyrhynchus.* (Cope). Fig. 392. Axillary process of pelvic fin well developed. Lower lip (Fig. 391B) moderately notched in center with 3-4 rows of papillae in mid-line; outer surface of upper lip without papillae. Length 7-8 inches. Widespread in Great Basin of Utah and Wyoming, Fraser and upper Columbia drainages; Green River in Utah and Wyoming; upper Saskatchewan River drainage; upper Missouri River drainage including the Black Hills.

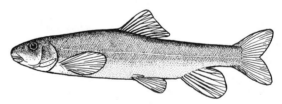

Figure 392.

RIO GRANDE SUCKER, *Catostomus (Pantosteus) plebius* Baird and Girard. Axillary process of pelvic fin undeveloped. Lower lip with deep median notch, only 2-3 rows of papillae in mid-line (Fig. 391C), outer face of upper lip with papillae. Length 4-6 inches. Rio Grande River drainage in Colorado, New Mexico and northern Mexico.

COLORADO SUCKER, *Catostomus (Pantosteus) discobolus* Cope. Fig. 393. Axillary process of pelvic fin undeveloped. Lower lip with shallow median notch, more than 3 rows of papillae in mid-line. Predorsal scales usually more than 50. Colorado River drainage above Grand Canyon, upper Snake River drainage and drainage of Bear River and Weber Lake, Idaho, Utah and Nevada. Length up to 18 inches. Quite variable and includes former species, *P. virescens* Cope, *P. delphinus* (Cope) and others.

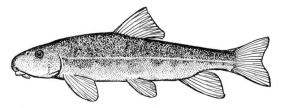

Figure 393.

DESERT SUCKER, *Catostomus (Pantosteus) clarki* Baird and Girard. Fig. 394. Axillary process of pelvic fin poorly developed or absent. Lower lip (Fig. 391A) with shallow median notch, 4-5 rows of papillae in midline, outer face of upper lip without papillae. Predorsal scale number variable, less than 50 (13-50). Length up to 14 inches. Colorado River drainage below Grand Canyon including Gila River system of Arizona and New Mexico, White River and Virgin River drainages in Nevada and Utah.

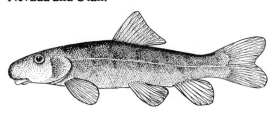

Figure 394.

SANTA ANA SUCKER, *Catostomus (Pantosteus) santaanae* (Snyder). Axillary process of pelvic fin variable in development. Lower lip with shallow notch and with numerous rows of papillae in mid-line. Outer face of upper lip may be weakly papillose. Length about 6 inches. Santa Clara, San Gabriel and Santa Ana River drainages in southern California.

BRIDGELIP SUCKER, *Catostomus (Pantosteus) columbianus* (Eigenmann and Eigenmann). Some individuals may show shallow notches between upper and lower lips and may key here. See couplet 14b.

13b Lower lip not separated by lateral notches from upper lip; both lips papillose. *Catostomus* Figs. 395A, B, C 14

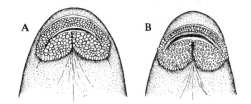

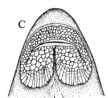

Figure 395.

14a Coarse scaled suckers, less than 80 lateral line scales. Fig. 396 . . . . WHITE SUCKER, *Catostomus commersoni* (Lacepede)

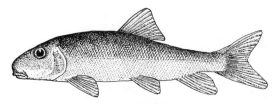

Figure 396.

Lateral line scales 55-70. Lips papillose and almost divided (Fig. 395A). Very common east of the Rockies from northwestern Canada eastward and to Colorado, Missouri and Georgia. It is represented on the Great Plains from Montana to Mexico by the sub-species *C. c. suckli* Girard. Introduced elsewhere including the Colorado River system. Length up to 18 inches.

UTAH SUCKER, *Catostomus ardens* Jordan and Gilbert. Lateral line scales 70-72. Upper lip deep, lower lip almost divided by deep median groove. Bonneville Basin, Utah and upper Snake River drainage in Wyoming, Idaho and Nevada.

WEBUG SUCKER, *Catostomus fecundus* Cope and Yarrow. Lateral line scales 64-75. Lower lip deeply incised. Utah Lake, Utah.

LARGESCALE SUCKER, *Catostomus macrocheilus* Girard. Lateral line scales 65-75. Upper lip wide; lower lip with deep median groove. Sides with dusky lateral band. Coastal drainage from Skeena River south into Oregon.

SONORA SUCKER, *Catostomus insignis* Baird and Girard. Lateral line scales 56-67. Gila River system, Arizona and New Mexico.

WARNER SUCKER, *Catostomus warnerensis* Snyder. Lower lip with shallow median groove. Lateral line scales 69-77. Warner Lakes drainage, Oregon.

KLAMATH LARGESCALE SUCKER, *Catostomus snyderi* Gilbert. Lateral line scales 69-77. Deep median groove in lower lip. Above falls of Klamath River, Oregon and California.

SACRAMENTO SUCKER, *Catostomus occidentalis* Ayres. Fig. 397. Lateral line scales 58-75; lips thin; dorsal rays 12-14. Four subspecies. Sacramento-San Joaquin Rivers and coastal streams northward from San Francisco Bay in California.

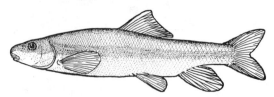

Figure 397.

14b Fine scaled suckers, more than 80 scales in lateral line. Fig. 398. . . . . LONGNOSE SUCKER, *Catostomus catostomus* **(Forster)**

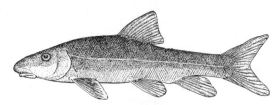

Figure 398.

Lateral line scales 90-117; snout long and pointed. Lower lip large and almost divided (Fig. 395B). Breeding males with distinct reddish lateral band. Length up to 18 inches. Widely distributed east of the Rockies from Alaska to Maine, in Great Lakes drainage but not in Mississippi River drainage except in drainage of upper Missouri River where it is represented by *C. c. griseus* Girard. Also present in Columbia River system west of the Rockies.

FLANNELMOUTH SUCKER, *Catostomus latipinnis* Baird and Girard. Fig. 399. Lower lip very thick and elongated (Fig. 395C), completely divided by median groove; dorsal rays 10-11. Lateral line scales more than 80. Length over 18 inches. Colorado River drainage.

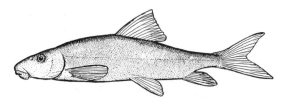

Figure 399.

MODOC SUCKER, *Catostomus microps* Rutter. Lateral line scales 81-87. Lower lip deeply divided; dorsal fin rays 10-11. Upper Pit River drainage, Modoc County, California.

TAHOE SUCKER, *Catostomus tahoensis* Gill and Jordan. Fig. 400. Lower lip deeply incised; lateral line scales 83-92. Breeding males bicolor, with prominent red lateral band. Lake

Tahoe and Lahontan Basin, California and Nevada.

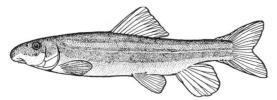

Figure 400.

OWENS SUCKER, *Catostomus fumiventris* Miller. Similar to Tahoe sucker but with larger scales, usually fewer than 80 scales in lateral line. Breeding males without prominent red stripe. Native to the Owens River and its tributaries, California.

KLAMATH SMALLSCALE SUCKER, *Catostomus rimiculus* Gilbert and Snyder. Fig. 401. Axillary process at base of pelvic fin not developed. Lower lip with deep median groove; lateral line scales 81-93. Trinity, Klamath and Rogue Rivers in Oregon and California.

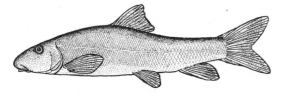

Figure 401.

BRIDGELIP SUCKER, *Catostomus columbianus* (Eigenmann and Eigenmann). Axillary process at base of pelvic fin not developed. Lateral line scales 100-111; lower lip not deeply notched; sometimes with weak lateral notches between upper and lower lips. See couplet 13a. Middle and lower Columbia River northward to Fraser River.

LOST RIVER SUCKER, *Catostomus luxatus* (Cope). Snout with distinct hump (Fig. 402). Lateral line scales more than 80 (82-88); lower lip with small papillae, a few on corners of upper lip (Fig. 403). Lower lip almost completely divided by wide median notch. Klamath River system in Oregon and California.

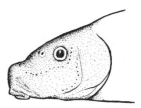

Figure 402.

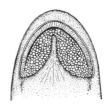

Figure 403.

CATFISH FAMILY
Ictaluridae

The members of this family are readily distinguished by their scaleless bodies, broad flat heads, sharp heavy pectoral and dorsal spines, and long barbels about the mouth. They possess bands of numerous bristle-like teeth in the upper jaw. Their barbels are arranged in a definite pattern, four under the jaws, two above and one on each tip of the maxillary. Originally the family Ictaluridae was found in the United States only east of the Rockies, but now various species have been widely introduced in the western states. All the larger species are desirable food fishes. This family includes the large catfishes, the bullheads, and the small madtoms. The latter are noted for venomous glands in their spines. All catfishes possess glands which causes a prick from their spines to be quite irritating, but the small madtoms have glands which seem to be more virulent.

These fishes are all more or less omnivorous, feeding on all manner of animal food and often on vegetable matter. They are mostly nocturnal and use their barbels to locate their food. They are nesting fishes, spawning in the spring or early summer and depositing their eggs in some sort of a depression or cavity. The male assumes care of the eggs and guards the young for several weeks after hatching.

The marine catfishes belong to another family and can be recognized by having fewer barbels. The gafftopsail catfish, *Bagre marinus* (Mitchill) and the sea catfish, *Arius felis* (Linnaeus), found in the seas from Cape Cod to Texas, may occasionally enter fresh water. These are bluish silver catfishes with deeply forked tails. The gafftopsail catfish has 4 barbels on the head, and the sea catfish has 6 barbels on the head. The marine catfishes do not build nests but the male carries the eggs in his mouth.

1a Caudal fin deeply forked 2

1b Caudal fin not deeply forked, but rounded or square 4

2a Anal fin with 30 or more rays. Fig. 404 . . . . **BLUE CATFISH,** *Ictalurus furcatus* (Lesueur)

Figure 404.

Silvery white below shading to dusky blue on back. Reaches weight of over 100 pounds. Large rivers from Minnesota and Ohio southward into Mexico. Rare if not extinct in the north. Introduced into southern California.

2b Anal fin with fewer than 30 rays 3

3a Anal fin with 19 to 23 rays. Fig. 405 . . . . **WHITE CATFISH,** *Ictalurus catus* (Linnaeus)

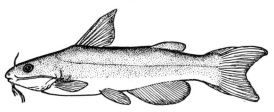

Figure 405.

Bluish above and silvery below. Reaches a length of 2 feet. Coastal streams from Chesapeake Bay region and southward to Texas; widely introduced on west coast.

SPOTTED BULLHEAD, *Ictalurus serracanthus* Yerger and Relyea. Resembles the white catfish but has small yellowish spots on the sides. It is present in the Apalachicola, Ochlockonee, and Suwannee River drainages of Florida, Georgia and Alabama.

3b Anal fin with 24-29 rays. Fig. 406 . . . . **CHANNEL CATFISH,** *Ictalurus punctatus* (Rafinesque)

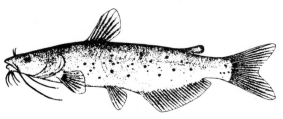

Figure 406.

Whitish below and on sides, bluish on back, sides with small irregular spots. Reaches a weight of over 20 pounds. Great Lakes and Saskatchewan River southward to Gulf and Mexico. Introduced elsewhere including Hawaii.

HEADWATER CATFISH, *Ictalurus lupus* (Girard). Similar to channel catfish with narrow head and slender body but lacks small spots on sides and caudal fin is not as deeply forked. Anal fin base is longer than head length; anal rays less than 25. Northeastern Mexico and reported from Pecos River drainage of Texas.

YAQUI CATFISH, *Ictalurus pricei* (Rutter). Similar to the headwater catfish. It has about 24 anal rays, the anal fin base is shorter than the head length. Anterior part of the dorsal fin conspicuously higher than posterior part. Caudal fin is not as deeply forked as in channel catfish. Northern Mexico into southern Arizona.

4a Adipose fin free, not fused to caudal fin 5

4b Adipose fin not free but fused to caudal fin or separated by a slight or incomplete notch 9

5a Anal rays less than 16; band of teeth in upper jaw with backward lateral extensions (Fig. 407A). Fig. 408 . . . . FLATHEAD CATFISH, *Pylodictis olivaris* (Rafinesque)

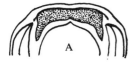

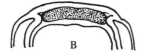

Figure 407.

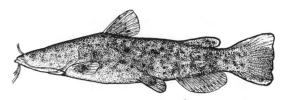

Figure 408.

Yellowish brown above, pale gray below and often mottled on sides. Reaches a weight of over 100 pounds. Large rivers, Mississippi valley into Mexico. Introduced into the lower Colorado River where they are now common.

5b Anal fin with more than 16 rays; band of teeth in upper jaw without any backward lateral extensions (Fig. 407B). (Formerly *Ameiurus,* BULLHEADS, but now part of *Ictalurus*) 6

6a Barbels under jaw white, not pigmented . . . . YELLOW BULLHEAD, *Ictalurus natalis* (Lesueur)

Color variable, back and sides various shades of brown to almost black; belly more or less yellow. Anal rays 23-27. Length up to 18 inches. North Dakota to Hudson River and south to Gulf. Introduced elsewhere.

6b Barbels under jaw gray to black, pigmented 7

7a Dark streak across base of dorsal fin; fins except pectoral edged with black. Fig. 409 . . . . FLAT BULLHEAD, *Ictalurus platycephalus* (Girard)

Figure 409.

Various shades of mottled olive brown to yellowish brown; head broad and flat, body elongated; barbels under jaw pale. Anal rays 21-24. Length up to 15 inches. Coastal streams, Roanoke River, Virginia to Altamaha River, Georgia.

GREEN BULLHEAD, *Ictalurus brunneus* (Jordan). Long considered as a form of the flat bullhead but differs in having fewer anal rays (17-20) and is seldom mottled (except in St. Johns River). Cape Fear River, North Carolina to St. Johns River, Florida.

7b No dark streak across base of dorsal fin; fins without black edges. 8

8a Pectoral spines strongly barbed on posterior edge (Fig. 410A), offers resistance when grasped by thumb and forefinger. Fig. 411 . . . .
. . . . BROWN BULLHEAD, *Ictalurus nebulosus* (Lesueur)

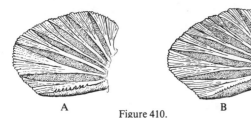

Figure 410.

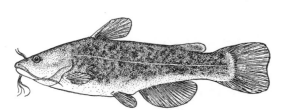

Figure 411.

Back and sides more or less dark brown, often mottled, belly gray to yellowish. Anal rays 20-30. Length to 18 inches. North Dakota and Saskatchewan to Nova Scotia and south to Mexico and Florida. Widely introduced elsewhere.

8b Pectoral spine weakly barbed on posterior edge (Fig. 410B), offers little resistance when grasped by thumb and forefinger. Fig. 412. . . . .
. . . . BLACK BULLHEAD, *Ictalurus melas* (Rafinesque)

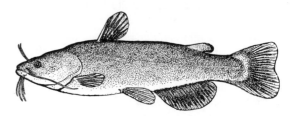

Figure 412.

Dark above, ranging from yellowish brown to almost black; belly varies from gray to yellowish. Anal rays 17-24. Length to 18 inches. North Dakota to New York and south to Texas, common throughout California.

9a Pectoral spine with serrae or barbs on the anterior and posterior margin (Fig. 413): serrae on anterior surface usually small, sometimes barely visible, those on the posterior surface large; pectoral spine curved; color pattern usually of dark blotches or saddle-like markings over lighter background 10

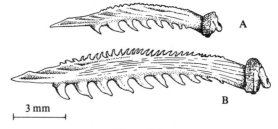

3 mm

Figure 413.

9b Pectoral spine without serrae on the anterior margin; posterior margin smooth, often with recurved hooks or step-like notches (Fig. 414); pectoral spine straight to moderately curved; dark color

pattern, without prominent blotches or saddle-like markings 20

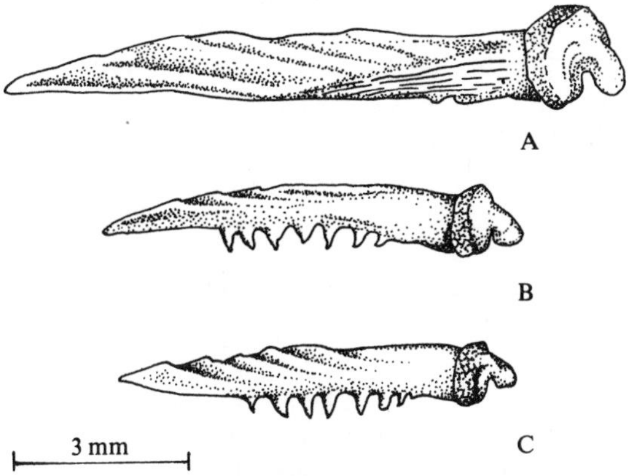

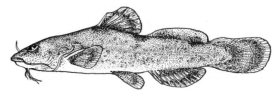

3 mm

Figure 414.

10a Pelvic fin rays typically 8 11

10b Pelvic fin rays typically 9 12

11a Pectoral fin soft rays typically 8 or 9 (7-10); anal fin rays usually 13 (12-17); mouth inferior, lower jaw included . . . . LEAST MADTOM, *Noturus hildebrandi* (Bailey and Taylor)
Background color yellowish, with 3 prominent saddle-like blotches alternating with light areas on the back and upper side. Small, slightly more than 2 inches in total length. Eastern tributaries of the Mississippi River in Tennessee, Mississippi and possibly Kentucky.

11b Pectoral fin soft rays 7 or 8; anal fin rays 12 or 13; mouth subterminal, lower jaw only slightly included. . . . . SMOKY MADTOM, *Noturus baileyi* Taylor
Medium brown color with small dorsal saddles. Caudal fin short. Abrams Creek, tributary of Little Tennessee River, Great Smoky Mountain National Park.

12a Adipose fin nearly separated from caudal fin, with a free posterior flap; soft pectoral fin rays usually 8. Figs. 415, 413A . . . . MOUNTAIN MADTOM, *Noturus eleutherus* Jordan

Figure 415.

Dark to yellowish brown in color with mottling on the side of body; a dark bar at caudal peduncle base. Length to 4 inches. Preopercular mandibular pores typically 10. A river species. The Ohio River and tributaries from French Creek, Pennsylvania, Ohio, Kentucky, to the Wabash River, Indiana. The Cumberland River in Tennessee, upper Tennessee River in North Carolina, Georgia and Tennessee. Disjunct population in Ouachita River of Arkansas and Red River system of Oklahoma and Arkansas.

CHECKERED MADTOM, *Noturus flavater* Taylor. Yellowish brown with brown or purplish black saddles and blotches. Broad black bar at caudal base and black blotch on outer third of dorsal fin. Preopercular mandibular pores typically 11. Total length to 5 inches. White River system in Arkansas and Missouri.

12b Adipose fin and caudal fin united, moderately to broadly connected; adipose fin without free posterior flap 13

13a Predorsal length usually more than 1.5 times (1.4-2.0) in the distance from the dorsal origin to the end of the adipose fin 14

13b Predorsal length usually less than 1.5 times (1.1-1.6) in the distance from the dorsal origin to the end of the adipose fin 15

14a **Prominent dark bar at caudal base; soft pectoral fin rays typically 9. Fig. 416** . . . . **OZARK MADTOM,** *Noturus albater* **Taylor**

Figure 416.

Color yellowish with black or brown blotches, a large dark adipose bar, and white blotch on dorsal base of caudal fin rays. Length to 3 or 4 inches. Moderate to large streams in the upper White and St. Francis systems of Arkansas and Missouri.

14b **Bar, if present, on caudal base not prominent nor darker than bands on the caudal fin; soft pectoral fin rays typically 8 (occasionally 9). Fig. 417** . . . . **CADDO MADTOM,** *Noturus taylori* **Douglas**

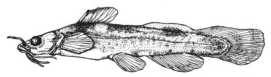

Figure 417.

Slender fish with prominent dark dorsal saddles; dorsal fin tipped with black; sub-marginal adipose fin blotch present. Anal fin rays 13 to 16, typically less than 16. Two to 3 inches in length. Upper Caddo River, a tributary of the Ouachita River, Arkansas.

ELEGANT MADTOM, *Noturus elegans* Taylor. Similar in many respects to the Caddo madtom but with 14 to 19 anal rays, usually 15 or more. Small creeks and small rivers in Duck

River system, Tennessee and Green River system, Kentucky.

SCIOTO MADTOM, *Noturus trautmani* Taylor. Similar to the preceding two species but differing in having a yellowish adipose fin without a black bar. Anal fin rays 13 to 16, typically 14. Known from a single locality, Big Darby Creek, Scioto River, Ohio.

15a **Dark bar on adipose fin always extending into upper half of adipose fin** 16

15b **Dusky bar on adipose fin not extending into upper half of adipose fin** . . . . **NEOSHO MADTOM,** *Noturus placidus* **Taylor**

Color light yellowish pink, mottled with brown; dorsal spine dusky yellowish white at tip; dorsal fin with dark blotch extending to base of second ray; pectoral fin with blotches of pigment. Predorsal light spots in the dorsal fin saddle. Length to 3 inches. Riffles in main channels of the Cottonwood and Neosho Rivers, Kansas, lower few miles of Illinois River, Oklahoma.

16a **Adipose blotch or band extending to fin margin.** 17

16b **Adipose blotch or band not extending to fin margin** 19

17a **Adipose fin nearly free from caudal fin; with 2 crescentic bands on caudal fin** . . . . **FRECKLEBELLY MADTOM,** *Noturus munitus* **Suttkus and Taylor**

Yellowish in color, heavily mottled with dark brown. With a dark bar posterior to head. Anal rays 12-14. Caudal rays 45-52. Riffles and rapids of the Pearl River. Louisiana and Mississippi, east to Cahaba River, a tributary of the Alabama River, Alabama.

17b Adipose fin and caudal fin broadly connected; no mid-caudal crescent present. 18

18a Dorsal fin with black blotch on extremity of anterior rays. Fig. 418 . . . .
. . . . **BRINDLED MADTOM,**
Noturus miurus Jordan

Figure 418.

Color varies from yellowish, light reddish orange to pinkish. Four dark saddles on back. Caudal fin grayish with black sub-marginal band. Caudal fin rays 57 or more (54-65). Length about 4 inches. Lowland streams with some current, in pools and lakes. Ohio River Valley, in tributaries of the Ohio River, south to Pearl River, Louisiana and Mississippi.

18b Dorsal fin yellowish without black blotch on margin . . . .
. . . . **YELLOWFIN MADTOM,**
Noturus flavipinnis Taylor
Brownish tan with four prominent dorsal saddles. Brownish patch on posterior end of caudal peduncle continues above and below as black band across caudal rays. Caudal fin yellowish with a broad brownish band near margin. Caudal fin rays 54 to 63. Less than four inches long. Pools and slow running water in tributaries of the upper Tennessee River basin, Georgia, Tennessee and Virginia. Clench River system in Virginia.

19a A pair of large light spots enclosed in a dark saddle just anterior to dorsal fin base; caudal rays usually 53 or fewer. . . .

. . . . **NORTHERN MADTOM,**
Noturus stigmosus Taylor
Color pinkish, yellowish or light tan, mottled with light brown. Caudal fin yellowish white with two crescentic bands. Blackish adipose blotch extends over half of fin. Typically in large rivers. Tributaries of Mississippi River in western Tennessee, Ohio River system from Pennsylvania through Ohio to Wabash River, Indiana; Green River in Kentucky. Lake Erie basin, Detroit, Huron and Maumee Rivers in Great Lakes basin.

19b Dark saddle at dorsal fin nearly uniform, without light spots; caudal fin rays usually 53 or more. Figs. 419, 413B . . . .
. . . . **CAROLINA MADTOM,**
Noturus furiosus Jordan and Meek

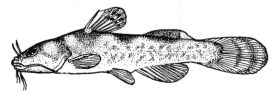

Figure 419.

Variegated body color, basidorsal saddle extends back to third dorsal ray and forward to mid-point between head and dorsal fin; dorsal fin with some pigment and sub-terminal brownish band. Caudal fin with 2 crescentic bands. Length about 5 inches. Neuse and Tar Rivers of North Carolina.

20a Teeth in upper jaw, premaxillary, in bands with posterior extensions. Figs. 407A, 420 . . . .
. . . . **STONECAT,** *Noturus flavus* **(Rafinesque)**

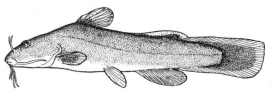

Figure 420.

Color yellowish brown to gray. Reaches a length of 12 inches. Montana to Great Lakes and south to Texas.

20b **Teeth in upper jaw, premaxillary, in bands without posterior lateral extensions. Fig. 407B** 21

21a **Mouth terminal, jaws typically of equal length** 22

21b **Mouth inferior, lower jaw typically included.** 24

22a **Posterior edge of pectoral fin lacking serrae.** 23

22b **Posterior edge of pectoral fin with obvious well-developed serrae. Figs. 414C, 421** . . . . **SLENDER MADTOM,** *Noturus exilis* **Nelson**

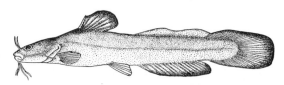

Figure 421.

Slender dark yellowish brown to gray-black. The dorsal, anal and caudal fin with dark borders. Anal rays 17 to 20; usually 9 pectoral rays. Distance from adipose fin to tip of caudal fin over 1.6 times in distance from dorsal origin to end of adipose fin. Reaches a length of

slightly more than 6 inches. Small creeks from southeastern Minnesota, southern Wisconsin, Iowa, Illinois, Missouri, eastern Kansas and the Ozarks; disjunct population in Tennessee and Cumberland River basins, in Kentucky, Tennessee and northern Alabama.

23a **Head short, 3.7 to 4.2 times in standard length; anal rays typically 17 or more (16-19).** . . . . **OUACHITA MADTOM,** *Noturus lachneri* **Taylor**

Moderately elongate, dark brown in color. Usually with 9 pectoral rays. Less than 4 inches in length. Saline River, a tributary to the Ouachita River, Arkansas.

23b **Head longer, 3.0 to 3.8 times in standard length; anal fin rays averaging less than 17 (12-18). Figs. 422, 414A.** . . . . **TADPOLE MADTOM,** *Noturus gyrinus* **(Mitchill)**

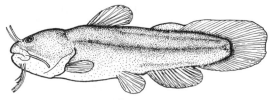

Figure 422.

Yellowish brown to gray in color. Usually with 7 or fewer pectoral fin rays. Reaches a length of 5 inches. A lake, pond or slow running water species. Widely distributed from Texas to Saskatchewan, including the Dakotas, Minnesota and east to the Lake Ontario drainage in New York. Atlantic coastal drainage below the Fall Line from the Hudson River to the Florida Gulf coast drainage west to Texas. Absent from the Appalachian Highlands and Piedmont Plateau. Introduced in Connecticut River, Massachusetts and Snake River of Idaho and Oregon.

24a Pelvic fin rays typically 8: body slightly mottled, giving the appearance of freckles. Fig. 423 . . . .
. . . . **SPECKLED MADTOM,** *Noturus leptacanthus* **Jordan**

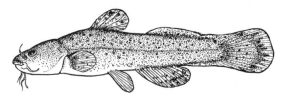

Figure 423.

Yellowish brown to reddish brown in color, with very large chromatophores scattered over the body and fins. The pectoral spine is short and smooth. Reaches a length of 3 inches. Atlantic and Gulf coastal drainages of southeastern United States; from eastern Louisiana, Mississippi, Georgia, and Florida to Edisto River, South Carolina.

24b Pelvic fin rays typically 9 or more; no mottling or clusters of pigment over body or fins 25

25a Anal fin typically with 20 or more rays; abdomen and ventral surface of head well pigmented; dorsal spine slender and flexible in young and juveniles 26

25b Anal fin typically with fewer than 20 rays; abdomen and ventral surface of head with little pigment; dorsal spine stout and stiff in all ages. 27

26a Anal fin typically with 21-24 rays; soft pectoral rays typically 9. Fig. 424 . . . . . .
. . . . **BLACK MADTOM,** *Noturus funebris* **Jordan and Swain**

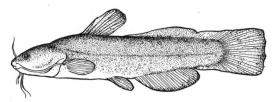

Figure 424.

Body rather darkish, ventral surface with speckling of pigment; fins dusky, dorsal, anal and caudal fins dark on margins. Pectoral spine with weak serrae or serrae absent. Creeks and small streams in Gulf coastal drainage, from Pearl River, Louisiana and Mississippi eastward to Econfina Creek, Florida.

26b Anal fin typically with 20-22 rays; soft pectoral rays 8 or 9. Figs. 425, 414B . . . . .
. . . . **BROWN MADTOM,** *Noturus phaeus* **Taylor**

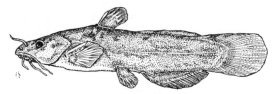

Figure 425.

Similar to the black madtom but differing in having well-developed serrae on the pectoral spine. Light to dark brownish in color, fins pigmented with light edges. Dorsal and anal fins with dark band on border. Lower Mississippi River Valley, permanent springs and small streams. Eastern tributaries of the Mississippi River in southwestern Kentucky to Louisiana; Ouachita and Red River to northwestern Louisiana, Bayou Teche, Louisiana.

27a Dorsal and anal fins with broad light border; ventral lobe of caudal fin black; dorsal and pectoral spines short; pectoral spine length more than 2 times in length of head. Fig. 426 . . . .
. . . . ORANGEFIN MADTOM, *Noturus gilberti* Jordan and Evermann

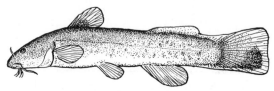

Figure 426.

Dark yellowish brown, ventral surface of head and abdomen immaculate. The adipose fin low, nearly free from caudal fin. Length to 4 inches. Rubble in fast water in the Mayo, James and Roanoke Rivers, Virginia.

27b Dorsal and anal fins dark, occasionally with a narrow light border; ventral lobe of caudal fin without black pigment or uniformly pigmented; dorsal and pectoral spines longer, pectoral spine length about 2 times in length of head 28

28a Pectoral spine lacking serrae, posterior edge slightly roughened; head rounded above. Fig. 427 . . . .
. . . . . FRECKLED MADTOM, *Noturus nocturnus* Jordan and Gilbert

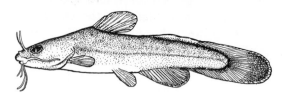

Figure 427.

Yellowish to dark brown, speckled with black or uniformly dusky. Fins dusky with a narrow light margin. Length to 3 inches. Lower and central Mississippi River drainage, Ohio River in Indiana and Kentucky; Missouri, Illinois, Kansas, Oklahoma and Texas; Tennessee River in Tennessee and Mississippi. Gulf coast drainage from Mobile River, Alabama west to San Jacinto River, Texas.

28b Pectoral spine with distinct serrae; head flattened above. Fig. 428 . . . .
. . . . MARGINED MADTOM, *Noturus insignis* (Richardson)

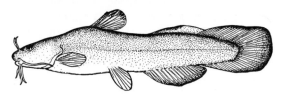

Figure 428.

Dusky colored. Fins tend to have blackish margins. Pectoral spine, posterior edge with nine serrae, these may become reduced in size in older individuals. Length to 5 or 6 inches. Lake Ontario drainage; coastal rivers from New York to Georgia. Upper Ohio River and Tennessee River, perhaps by introduction.

AIRBREATHING CATFISH FAMILY
Clariidae

Airbreathing catfishes have accessory air-breathing respiratory organs, the epibranchial organs, that extend backward from the gill or branchial chambers. These organs are highly vascularized and enable the fish to breathe air when out of water and to live in aquatic habitats devoid of oxygen. Members of the family are native to Africa and southern and southeastern Asia as far east as the Phillipines. The walking catfish, *Clarias batrachus* (Linnaeus), was apparently introduced accidentally from aquarists' holding ponds in southern Florida and also in the lower Colorado River. Since its initial release in Florida it has spread rapidly and federal officials have expressed concern that it will spread throughout the Gulf coast states west to Texas.

The walking catfish can move long distances over land on its stiffened pectoral and pelvic fins. Lateral body movements of the slender body aid in terrestrial locomotion. Extensive migrations usually take place on damp or rainy nights. The clariids are also tolerant to relatively high salinities so they can move freely from stream to stream along the coastal regions. They are voracious omnivores and compete with and feed on our native fishes. In Asia airbreathing catfishes are an important food resource but the walking catfish is not a food fish.

Both albinos and pigmented individuals are found in Florida waters. The clariids can be distinguished from our native catfishes by the presence of 4 pairs of very long filamentous barbels, a long eel-like body and a long dorsal fin. One pair of barbels is maxillary, one pair nasal and two pairs are on the mandible, Fig. 428A.

Figure 428A.

ARMORED CATFISH FAMILY
Loricariidae

The loricariid catfishes are native to South America and are favorites among fish fanciers. They are characterized by bony scutes covering the body much like the armor of medieval knights and the presence of 10 pairs of barbels surrounding the mouth. These catfishes live on the bottom in swift flowing streams and use their sucker-like mouths to cling to the substrate and also to feed.

One species, *Hypostomus plecostomus* (Linnaeus), native to Central America has been reported from a pond in West Miami, Florida and there are indications that other species may have been accidentally introduced in the same area. There are more tropical fish wholesalers and breeders in Florida than in any other state and it is not too surprising to find that the state has such a large number of exotic fish species. Two different species of armored catfish are known in the western United States, one in a warm spring in southern Nevada and another in the San Antonio River, Texas. What the impact of such introductions may have on the native fish fauna is unknown.

CAVEFISH FAMILY
Amblyopsidae

Although the members of this family are known as cave or blind fishes, several have functional eyes and may be found in springs and swamps. They are usually pale and rather colorless fishes which appear to be naked as their scales are minute and imbedded. They have flat naked heads and have many papilose sensory structures on their heads and bodies. Their eyes are degenerate or are poorly developed. The pelvic fins are either absent or are small. In the adults the anus moves anteriorly and is located in the throat (jugular). The females retain the fertilized eggs and give birth to living young.

Members of this family are usually found in underground streams of the limestone regions of Kentucky, Missouri, Arkansas, southern Illinois, southern Indiana, Tennessee, and northern Alabama. One species occurs in the swamps of the southeastern United States.

1a Sensory papillae in two or three rows on each half of the caudal fin 2

1b Sensory papillae absent or restricted to one row on each half of the caudal fin . . 3

2a Pelvic fins present, but small. Fig. 429 . . .
. . . . NORTHERN CAVEFISH,
Amblyopsis spelaea DeKay

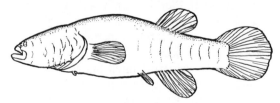

Figure 429.

Colorless; eyes absent or rudimentary; pelvic fins small. Numerous cross ridges (sensory) on body and head. Up to 5 inches length. Sub-terranean streams, caves in southern Indiana and in Mammouth Cave region of Kentucky.

2b Pelvic fins absent. . . . .
. . . . OZARK CAVEFISH,
Amblyopsis rosae (Eigenmann)
Similar in appearance to the northern cavefish, found in sub-terranean streams in southwestern Missouri. The Alabama cavefish, *Speoplatyrhinus poulsoni* Cooper and Kuehne, displays some similarities with the species of *Amblyopsis*. It differs in having a terminal mouth, no bifurcation of the caudal rays an extremely elongate head with a laterally constricted snout. The species is known from Key Cave, northwestern Alabama.

3a Eyes rudimentary and concealed; body colorless . . . .
. . . . SOUTHERN CAVEFISH,
Typhlichthys subterraneus Girard
Pelvic fins absent; body with numerous sensory ridges. Length 2 inches. Sub-terranean streams, Ozark region of Missouri and northeastern Oklahoma; and Kentucky through central Tennessee to northern Alabama.

3b Eyes developed and not concealed; body dark, pigment present in skin. 4

4a Body with ridges of papillae; caudal fin dark brown with several vertical rows of white specks or blotches. Fig. 430 . . . . . .
. . . . SPRING CAVEFISH,
Chologaster agassizi Putnam

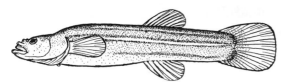
Figure 430.

Brown above, lighter below, entire body sprinkled with black specks. Three dark longitudinal stripes on sides. Center stripe may become pale. Length 1 1/2 inches. Springs in southern Illinois, Kentucky and Tennessee. A population of the spring cavefish has recently been discovered in southeastern Missouri.

4b **Body without ridges or papillae; caudal fin with black blotch at base, behind which is a white blotch or 2 white spots (may form a bar); remainder of fin is black . . . . SWAMPFISH,** *Chologaster cornuta* **Agassiz**

Dark brown above, whitish below, entire body sprinkled with black specks. Three black longitudinal stripes on sides. Dorsal fin white, may be edged with black. Length up to 2 inches. Found in lowland swamps from southern Virginia to central Georgia.

PIRATE PERCH FAMILY
Aphredoderidae

This family contains but one species, the pirate perch, *Aphredoderus sayanus* (Gilliams), Fig. 431, in the central Mississippi valley from southern Minnesota southward and along the coastal plain from New York southward to eastern Texas.

The body is dark olive, somewhat speckled; with two dark bars at the base of the caudal fin and becomes quite irridescent during the spawning season. Length about 5 inches. The outstanding character is the location of the anus. This moves forward as the fish grows until it is located under the throat (jugular) of the adults. The lateral line on the sides is only slightly or partly developed in most mid-western specimens but those from the Atlantic coastal region show a much better developed lateral line.

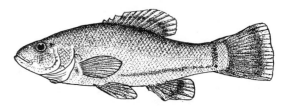
Figure 431.

The pirate perch build a nest and both parents are claimed to guard the nest. They are predaceous, feeding mostly on aquatic insects and other small aquatic animals.

TROUT-PERCH FAMILY
Percopsidae

These are small perch-like fishes with spiny fin rays but bearing a trout-like adipose fin. Only two species are in this family, one species being found in larger lakes and streams in the Great Lakes region and west to North Dakota and northward. Another species occurs in the Columbia River. They feed on small crustacea, aquatic insects, and other small aquatic animals.

1a One very weak spine in anal fin; 2 thin and weak spines in dorsal fin. Fig. 432. . . . . .TROUT-PERCH, Percopsis omiscomaycus (Walbaum)

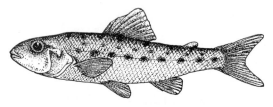

Figure 432.

Greenish yellow or straw color and mottled with a row of spots on the lateral line above which is another row of spots. Length 6 to 8 inches. Great Lakes region and upper Mississippi drainage northward into Yukon Territory.

1b Two very stout spines in anal fin; 2 stout spines in dorsal fin. Fig. 433 . . . . SAND ROLLER, Columbia transmontana (Eigenmann and Eigenmann)

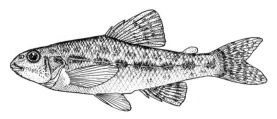

Figure 433.

Greenish yellow, sides mottled with numerous spots above and on the lateral line. Length about 6 inches. Lower Columbia River drainage.

COD FAMILY
Gadidae

The Cod family is an important marine family with several species entering or living in fresh water. The members of this family are characterized by a long soft dorsal fin which in some species may be divided into 2 or 3 parts. They bear a single barbel under the chin and possess small pelvic fins which are located under the throat (jugular).

The burbot, *Lota lota* (Linnaeus) (Fig. 434), is the only strictly freshwater species of cod found in North America. It is grayish olive and highly mottled with very minute scales. It reaches a length of 30 inches and a weight of over 10 pounds. The burbot is commonly found in larger lakes and often in very deep water. It spawns in the middle of winter

Figure 434.

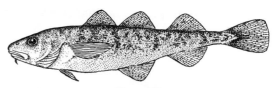

Figure 435.

either in shallow water or in streams. It ranges through the Great Lakes region and extreme northern Mississippi drainage of Wisconsin and Minnesota, and the upper Missouri River drainage, northwestward into Alaska and Siberia.

The tomcod, *Microgadus tomcod*

(Walbaum) (Fig. 435), is an inshore marine cod commonly entering estuaries to spawn in the winter. This is a mottled olive brown fish with the dorsal fin divided into three parts. It reaches a length of 12 inches. The tomcod ranges from Laborador to Virginia.

NEEDLEFISH FAMILY
Belonidae

This is a family of very slender marine fishes with both jaws elongated like a beak and bearing a superficial resemblance to the gars. The

Figure 436.

Atlantic needlefish, *Strongylura marina* (Walbaum) (Fig. 436), enters fresh water and ascends rivers from Cape Cod to Texas. This is a greenish silvery fish with a narrow lateral band, reaching a length of four feet although most specimens caught are much smaller.

KILLIFISH FAMILY
Cyprinodontidae

The killifishes and topminnows are small fishes found in fresh and salt water. Some are deep bodied, and others are quite slender. Many species show strong differences in the color and markings of the sexes.

They possess more or less protruding lower jaws and tilted mouths which are well adapted for surface feeding. Their lateral line is incomplete or only partially developed.

Many species live in the sea, and some of

these may enter fresh water. A number of other species are restricted to fresh water. Several species are isolated in springs of the deserts of the southwestern United States and have become highly modified, some have lost their pelvic fins and others have had their pelvic fins reduced in size. Killifishes feed on small crustacea and other aquatic animals. They spawn in the spring or early summer.

1a Teeth incisor-like and notched, with 2 or 3 cusps. Fig. 437 .... **2**

Figure 437.

1b Teeth pointed or conical, not with several cusps **5**

2a Teeth with 2 cusps .... *Crenichthys* Olivaceous fishes restricted to several desert streams.

RAILROAD VALLEY KILLIFISH, *Crenichthys nevadae* Hubbs. Fig. 438. Single row of spots on sides. Railroad Valley, Nevada.

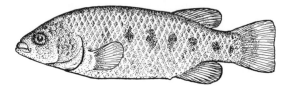

Figure 438.

WHITE RIVER KILLIFISH, *Crenichthys baileyi* (Gilbert). Fig. 439. Double row of spots on sides which may be connected. Moapa River, California, and White River and Pahranagat Valleys, Nevada.

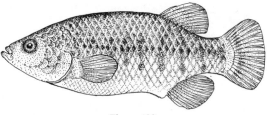

Figure 439.

2b Teeth with 3 cusps **3**

3a Dorsal fin long, 16-18 rays; first ray a stout grooved spine .... *Jordanella* FLAGFISH, *Jordanella floridae* Goode and Bean. Fig. 440. A deep bodied fish, olivaceous with orange or brassy sides and 4-5 diffuse dark crossbars. Gill membrane joined to shoulder a short distance above base of pectoral fin. Each series of scales form a broad longitudinal stripe; large diffuse spot on side below front of dorsal fin. Fins dusky, may be speckled or barred, and may have a spot in posterior part of dorsal fin. Length 2 1/2 inches. Coastal swamps and lagoons from Florida to Yucatan.

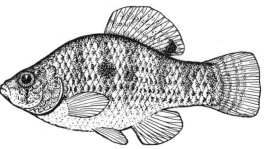

Figure 440.

3b Dorsal fin short, 10-12 rays; first dorsal ray not spine-like, but usually slender and rudimentary **4**

4a Opercular opening closed above by union of gill membrane to shoulder just above base of pectoral fin .... *Cyprinodon*

SHEEPSHEAD MINNOW, *Cyprinodon variegatus* Lacepède. Fig. 441. Deep bodied; very large humeral scale; dorsal rays 11; anal rays 10. Olivaceous; males deeply lustrous with salmon color belly; dorsal fin blackish with orange margin; caudal fin olive with black margin; anal and pelvic fins dusky with orange margins. Females lighter and with numerous crossbars. Young have black spots in dorsal fin near posterior tip. Length 2-3 inches. Widespread in fresh and brackish water along the coast from Cape Cod to the Rio Grande River.

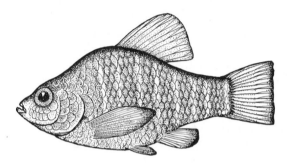

Figure 441.

LAKE EUSTIS MINNOW, *Cyprinodon hubbsi* Carr. Closely related to sheepshead minnow but with bars on sides and more rounded caudal fin. Lake Eustis area, Florida.

RED RIVER PUPFISH, *Cyprinodon rubrofluviatilis* Fowler. Fig. 442. Rather slender; body depth about three times in length; dorsal rays 9; anal rays 8-9; belly naked. Length 1 1/2 inches. Western Texas and southwestern Oklahoma.

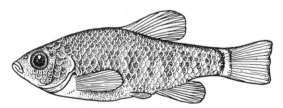

Figure 442.

LEON SPRING PUPFISH, *Cyprinodon bovinus* Baird and Girard, similar to the Red River pupfish, was rediscovered in Leon Creek below Leon Springs in southeastern Texas recently by Echelle and Miller. It was thought to be extinct and is no longer present in Leon Springs.

WHITE SANDS PUPFISH, *Cyprinodon tularosa* Miller and Echelle. Pelvic fin rays 6 or 7 (3-7); anal fin rays 9 or 10 (9-11). Scales in the lateral series 26-28. Breeding males yellow or orangish. Salt Creek and Malpais Spring, Tularosa Basin, New Mexico.

DESERT PUPFISH, *Cyprinodon macularius* Baird and Girard. Fig. 443. Variable in color, usually dusky with light underparts; fins usually margined with black. Females with dark crossbars on sides. Dorsal fin may have a blotch posteriorly. Humeral scale not much enlarged; dorsal rays 9-11; and anal rays 10-11. Desert streams of southeastern California and several areas of Arizona. Many related species in desert water holes as follows:

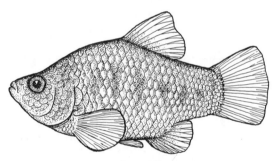

Figure 443.

ARMAGOSA PUPFISH, *Cyprinodon nevadensis* Eigenmann and Eigenmann. Pelvic fin rays 6 (0-9), sometimes missing. Scales in the lateral series 25-26 (23-28); predorsal scales 16-18 (15-24). Anal fin rays 10 (8-11). Six sub-species, represented by isolated populations in the Armagosa River Basin in California and Nevada.

DEVILS HOLE PUPFISH, *Cyprinodon diabolis* Wales. Smallest of all pupfish; pelvic fins absent. Devils Hole, Ash Meadows, Nye County, Nevada.

SALT CREEK PUPFISH, *Cyprinoon salinus* Miller. More slender than most pupfish. Pelvic fins small, sometimes absent, usually with 6 rays. Anal fin with 10 rays (9-11). Breeding males bluish with 5-8 broad vertical bands. Salt Creek, Death Valley, California.

COTTONBALL MARSH PUPFISH, *Cyprinodon milleri* LaBounty and Deacon. Scales in the lateral series 27-34; predorsal scales 26-33. Pelvic fin rays usually less than 6; anal rays 9-11. Coloration similar to Salt Creek pupfish. Cottonball Marsh, Death Valley, California.

OWENS PUPFISH, *Cyprinodon radiosus* Miller. Scales in the lateral series 26-27. Pelvic fin rays usually 7 (6-8); anal fin rays usually 10 (9-12). Breeding males bright blue with purplish lateral bars. Originally in Owens Valley, California, now restricted to Owens Valley Native Fish Sanctuary and Warm Springs, Lone Pine, Mono County, California.

4b Opercular opening closed by union of gill membrane to shoulder some distance above pectoral fin base about halfway to upper corner of opercle . . . .
. . . . . *Floridichthys*
GOLDSPOTTED KILLIFISH, *Floridichthys carpio* (Gunther). Fig. 444. Light olive. Males with silvery sides and about 6 faint coppery crossbars; sides sprinkled with yellow spots, dorsal and anal fins finely speckled and margined with orange. Female with plain fins and with many yellow blotches. Humeral scale not enlarged. Length 3 inches. Brackish water from Key West to Texas.

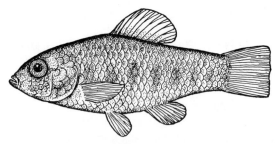

Figure 444.

5a Pelvic fins absent or undeveloped . . . . . .
. . . . . *Empetrichthys*
Dark brown above, lighter below and often somewhat mottled; fins dusky and may be speckled.

ASH MEADOWS KILLIFISH, *Empetrichthys merriami* Gilbert. Ash Meadows, Nye County, Nevada.

PAHRUMP KILLIFISH, *Empetrichthys latos* Miller. Fig. 445. Desert water holes in Pahrump Valley, Nye County, Nevada.

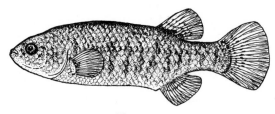

Figure 445.

5b Pelvic fins present 6

6a Teeth in more than one row; teeth in outer row may be large, and teeth in inner row or rows may be small 7

6b Teeth in a single row 8

7a Less than 30 scales in body length; upper margin of gill membrane (opercle) joined to shoulder just above base of pectoral fin . . . . *Adinia*

DIAMOND KILLIFISH, *Adinia xenica* (Jordan and Gilbert). Fig. 446. Deep bodied; scales in body length 25-28; more or less greenish with numerous crossbars; front of dorsal fin before front of pelvics. Length 2 inches. Brackish water from Florida to Texas. May enter fresh water.

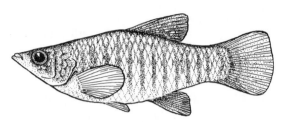

Figure 446.

RIVULUS, *Rivulus marmoratus* Poey. A small fish native to Cuba but has been reported from southern Florida. Body slender, head broad but with more than 30 scales. Body dashed with light and dark spots. Male with black spot on base of caudal fin; female with black spot on opercle.

7b More than 30 scales in body length; margin of gill membrane (opercle) not joined to shoulder just above base of pectoral fin but normal. . . . .*Fundulus*
Many species in this genus found in both salt and fresh water. See page 140 for key to species.

8a Body short, depth goes 3 1/4 to 3 3/4 times in length . . . .*Lucania*
RAINWATER KILLIFISH, *Lucania parva* (Baird and Girard). Fig. 447. Grayish and rather pale. Males with a large black spot at base of front of dusky orange dorsal fin; caudal fin edged with black. Females without dark edges on fins. Length 1 1/2 inches. Swamps and brackish waters along Atlantic coast from Cape Cod southward to Mexico. Isolated population in Pecos River of Texas and New

Mexico. Introduced into Irving Lake, southern California, established in streams flowing into San Francisco and Yaquina Bays, California, and Timpie Springs in Utah.

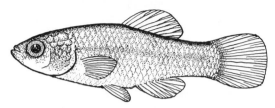

Figure 447.

8b Body more elongate, depth goes 4 1/4 to 5 times in length. 9

9a Body with black lateral band extending through eye to snout; dorsal rays 9; anal rays 9.
BLUEFIN KILLIFISH, *Lucania goodei* Jordan. Fig. 448. Olivaceous with black lateral band ending in caudal spot; black ventral band from vent to base of caudal fin. Male with basal half of dorsal and anal fins black, outer half pale with black margin. Swamps of Georgia and Florida.

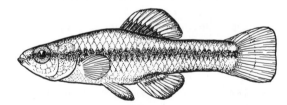

Figure 448.

9b Body without any black lateral band; mouth short and nearly vertical; dorsal rays 7, anal rays 8 . . . .*Leptolucania*
PYGMY KILLIFISH, *Leptolucania ommata* (Jordan). Fig. 449. Straw color. Male with 5-6 dark crossbars on sides. Female with black spot large as pupil on side just in front of anal fin. Large spot at upper part of base of caudal fin. Length 1 inch. Swamps of Georgia and Florida.

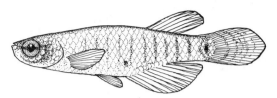

Figure 449.

KILLIFISH *GENUS FUNDULUS*

1a Front of dorsal fin above or before front of anal fin, never behind second ray of anal fin; dorsal fin with 10-17 rays 2

1b Front of dorsal fin behind second ray of anal fin; dorsal fin usually with 7-11 rays 10

2a Front of dorsal fin distinctly before front of anal fin 3

2b Front of dorsal fin above or slightly behind front of anal fin 8

3a Scales in body length 31-38; anal rays 10-12; body with or without crossbars . . 4

3b Scales in body length 40-60; anal rays 11-14; body with many crossbars 6

4a Female (Fig. 450) with 2 or 3 more or less broken black horizontal stripes; male (Fig. 451) with about 12 dark crossbars . . . . STRIPED KILLIFISH, *Fundulus majalis* (Walbaum)

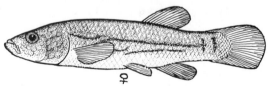

Figure 450.

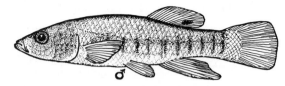

Figure 451.

Olivaceous above, pale below; black spot in posterior part of dorsal fin. Length 6 inches. Brackish water, sometimes entering fresh water, Cape Cod to Florida.

4b Female plain or with crossbars only, no longitudinal stripes 5

5a Males with dark crossbars. Fig. 452 . . . . LONGNOSE KILLIFISH, *Fundulus similis* (Baird and Girard)

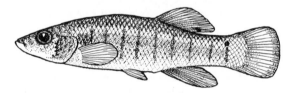

Figure 452.

Males and females with 10 to 15 crossbars. Olivaceous above, pale below. Males with large diffuse spot behind opercle. Young males have black spot in posterior part of dorsal fin. Both sexes with one or two black spots (diffuse in adults) near base of caudal fin. Length 6 inches. Brackish water, may enter fresh water, Florida to Texas.

CALIFORNIA KILLIFISH, *Fundulus parvipinnis* Girard. Fig. 453. Males with about 20 crossbars. Females with obscure lateral shade posteriorly. Pale olive green, somewhat mottled above, pale below. Length 4 inches. Brackish water, southern California.

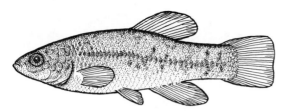

Figure 453.

5b Males with variable number of pale or silvery crossbars; females may be plain or may have about 15 dark crossbars; dorsal rays 11. . . . . MUMMICHOG, *Fundulus heteroclitus* **(Linnaeus)**

Dull green above, pale or yellowish below. Males (Fig. 454) with numerous white or yellow spots on sides; median fins dark; dorsal fin may have black blotch on last ray. Females

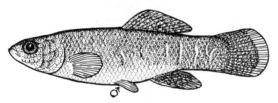

Figure 454.

(Fig. 455) with median fins plain. Length 5-6 inches. Brackish water, sometimes entering fresh water from Maine to eastern part of Gulf of Mexico.

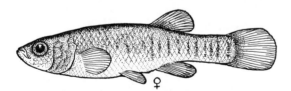

Figure 455.

GULF KILLIFISH, *Fundulus grandis* Baird and Girard. Fig. 456. Very similar to the mummichog, but differs largely in having smaller and shorter fins. Length 6 inches. Marine and brackish water, Florida to Texas.

Figure 456.

6a Scales in body length 44-48; anal rays 11. Fig. 457. . . . . BANDED KILLIFISH, *Fundulus diaphanus* **(Lesueur)**

Figure 457.

Olivaceous above with silvery sides, marked with about 20 more or less dark crossbars. Scale rows 41 to 52. Length up to 4 inches. Several sub-species widely distributed from North Dakota and Iowa to Quebec and South Carolina.

WACCAMAW KILLIFISH, *Fundulus waccamensis* Hubbs and Raney. Very similar but has a more slender body and more than 52 scale rows. Lake Waccamaw, North Carolina.

6b Scales in body length over 50; anal rays over 11 7

7a Dorsal fin rays more than 15 . . . . SEMINOLE KILLIFISH, *Fundulus seminolis* **Girard**

Olive green or yellowish brown. Males with longitudinal streaks formed by spots on scales; dorsal and caudal fins with large dark spots on bars; outer part of caudal mostly black. Females (Fig. 458) and young with 12-14 rather faint crossbars. Length 3 inches. Florida.

Figure 458.

**7b Dorsal fin rays less than 15. Fig. 459 . . . .
. . . . PLAINS KILLIFISH,
Fundulus kansae Garman**

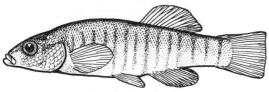

Figure 459.

Greenish with silvery sides and belly; sides marked with numerous crossbars (14-20); black spot in front of dorsal fin. Scale rows 50 or more. South Dakota and Wyoming south to Texas.

RIO GRANDE KILLIFISH, *Fundulus zebrinus* Jordan and Gilbert. Very similar and may be a sub-species. Scale rows less than 50. Trans Pecos region of Texas and New Mexico.

**8a Dorsal fin rays 13-15; anal fin rays 13-15
. . . . NORTHERN STUDFISH,
Fundulus catenatus (Storer)**

Bluish or green above and pale below with orange (male, Fig. 460) or brown (female, Fig. 461) spot on each scale; dorsal rays 14; anal rays 14-15. Length 6-7 inches. Highlands of eastern Tennessee and the Ozarks of Missouri and Arkansas.

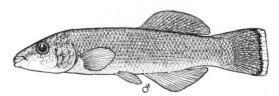

Figure 460.

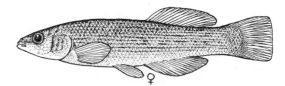

Figure 461.

SOUTHERN STUDFISH, *Fundulus stellifer* (Jordan). Fig. 462. Male bright blue above, silvery below; body and cheeks with irregularly scattered orange spots. Female with irregular brown dashes. Dorsal rays 13; anal rays 13. Length 4 inches. Alabama River system in Georgia and Alabama.

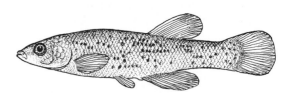

Figure 462.

8b Dorsal fin rays 10-11; anal fin rays 7-11 9

**9a Scales in body length less than 40. . . . . . .
. . . . MARSH KILLIFISH, *Fundulus confluentus* Goode and Bean**

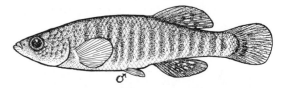

Figure 463.

Brownish yellow. Male (Fig. 463) with many crossbars. Female (Fig. 464) with sides marked with irregular dots; scales each with a dash appearing as numerous lateral streaks. Dorsal fin with black spot posteriorly; anal rays 7-9. Length 2 1/2 inches. Coastal swamps Maryland to Florida and west to Louisiana.

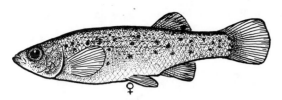

Figure 464.

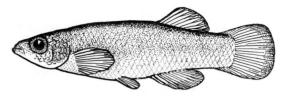

Figure 465.

BAYOU KILLIFISH, *Fundulus pulvereus* (Evermann). Very similar if not the same as the marsh killifish. Brackish water, Mobile, Alabama to Corpus Christi, Texas.

SALTMARSH TOPMINNOW, *Fundulus jenkinsi* (Evermann). Pale olivaceous with minute brownish specks except on breast; no distinct crossbars on sides. Males with 15-30 larger spots more or less in 2 rows on sides which may form indistinct crossbars. Brackish water, Gulf of Mexico, may enter fresh water.

SPECKLED KILLIFISH, *Fundulus rathbuni* Jordan and Meek. Similar to blackspot killifish. Pale green with irregular dark spots over body; anal rays 11. Young have pale crossbars. Length 2 1/2 inches. Streams of eastern North Carolina.

9b Scales in body length about 42.........
........WHITELINE TOPMINNOW,
Fundulus albolineatus **Gilbert**
Males dark brown with plumbeus sides, scale spots form interrupted whitish streaks on sides. Females olivaceous and silvery below. Scale rows form narrow black streaks. Length 3 1/2 inches. Tennessee River system.

10a Body either plain or with irregular spots. Fig. 465.........................
............PLAINS TOPMINNOW,
Fundulus sciadicus **Cope**

Uniformly olivaceous punctulated with fine dots, belly pale; no bars or streaks. Length 2 1/2 inches. Missouri River system, South Dakota to Colorado.

10b Body marked with either crossbars or longitudinal stripes 11

11a Body marked with single longitudinal stripe and random dots; no crossbars; female rather plain. Fig. 466 ....
...... BLACKSTRIPE TOPMINNOW,
Fundulus notatus **(Rafinesque)**

Figure 466.

Brownish green with broad lateral band from snout to base of caudal fin; spots on body and fins rather diffuse. Length 3 1/2 inches. Iowa to Ohio and south to parts of Tennessee, Mississippi, and central Texas.

BLACKSPOTTED TOPMINNOW, *Fundulus olivaceus* (Storer). Fig. 467. Differs

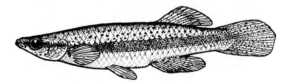

Figure 467.

slightly from the preceding. Spots on body are smaller and more concise. The predorsal and postdorsal stripe tends to be lacking in adults although present at least as a row of spots in the preceding. Length 3 1/2 inches. Oklahoma and Mississippi to western Florida and eastern Texas.

11b   Body marked with more than one longitudinal stripe or with crossbars. . . 12

12a   Male with dark bar below eye . . . .
. . . . STARHEAD TOPMINNOW,
***Fundulus nottii* (Agassiz)**
Pale olive. Male (Fig. 468) with faint longitudinal streaks formed by dots on each

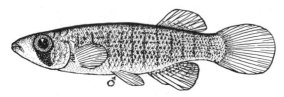

Figure 468.

scale, and about 10-12 crossbars. Female with longitudinal streaks, but crossbars faint or absent. Length 2 1/2 inches. Two sub-species; *F. n. nottii* (Agassiz) found in the Gulf coast drainage from western Florida, Mississippi and Louisiana east of the Mississippi River, and a northern sub-species *F. n. dispar* (Agassiz) in the Mississippi Valley from Iowa, southern Wisconsin, western Tennessee, Kentucky, and southeastern Arkansas; the Lake Michigan drainage of Michigan and Indiana. These are recognized as distinct species by some workers.
LINED TOPMINNOW, *Fundulus lineolatus* (Agassiz) (Fig. 469), has recently been recognized as a distinct species. It is similar in many respects to the starhead topminnow but has more orange on the top of the head and the sub-occular bar is not as high and diffuse.

South Carolina and Florida east of the Ochlochonee River.

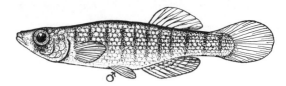

Figure 469.

BLAIR'S STARHEAD TOPMINNOW, *Fundulus blairae* Wiley and Hall. Similar to the preceding species but lacks vertical bars on the side of the body, dots on the side of body reddish in life but brown in preserved material. Eastern Texas, Brazos River, Galveston Bay drainage, southeastern Oklahoma, and Louisiana west of the Mississippi River.

12b   Male without dark bar below eye 13

13a   Crossbars of male usually less than 12 . . .
. . . . GOLDEN TOPMINNOW,
***Fundulus chrysotus* (Günther)**
Olivaceous above, light below and flecked with gold or orange, dorsal fin sometimes orange. Male (Fig. 470) with 6-10 crossbars; median

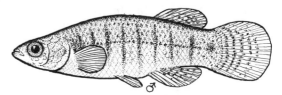

Figure 470.

fins speckled. Female (Fig. 471) more or less plain. Length 2 1/2 inches. South Carolina to Florida, Louisiana and Oklahoma.

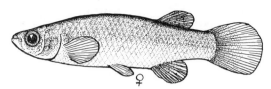

Figure 471.

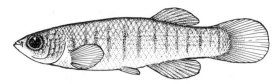

Figure 472.

13b Crossbars of both sexes usually more than 12. Fig. 472 . . . . BANDED TOPMINNOW, *Fundulus cingulatus* **Valenciennes**

Olivaceous. Male may have a few faint longitudinal streaks; body marked with 11-15 dark crossbars; belly orange and fins red. Females are more plain. Length 2 1/2 inches. Coastal swamps and streams, North Carolina, Florida and Alabama.

TOPMINNOW OR LIVEBEARER FAMILY
Poeciliidae

The members of this family are small fishes restricted to the southern part of the United States and farther south. They are closely allied to the killifish family and are hard to separate on structural characters.

They differ from most other freshwater fishes of the United States in their mode of reproduction, the females giving birth to living young. The males bear an intromittent organ, the *gonopodium,* developed from the modified anal fin. Fertilization is internal, the male depositing sperm in the genital tract of the female. The female carries the developing eggs until they hatch internally, and the young emerge alive from the female.

They feed on minute insects and other small animal forms. They are usually found in shallow sloughs and pools. Many species are popular for aquaria.

1a Origin of dorsal fin behind origin of anal fin; dorsal rays less than 12 2

1b Origin of dorsal fin over or in front of the origin of the anal fin; dorsal fin long and high (very high in males of some species); dorsal rays more than 12 . . . . *Poecilia*

SAILFIN MOLLY, *Poecilia latipinna* (Lesueur). Fig. 473. Light olive green above and lighter below, somewhat spotted; each scale with a spot. Dorsal fin marked with rows of spots; caudal fin with entire black margin. Length about 3 inches. Common near the coast from South Carolina into Mexico, abundant in southern California. Introduced in Hawaii.

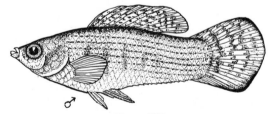

Figure 473.

AMAZON MOLLY, *Poecilia formosa* (Girard). Fig. 474. Scales with dark margins but with no spots. Front of dorsal fin farther back from head, dorsal fin with 12 or less rays. A peculiar fish known mostly as females, mating with males of related species although a few true males have been reported. Southern Texas into Mexico.

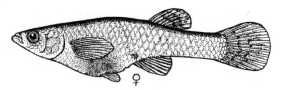

Figure 476.

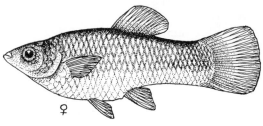

Figure 474.

2a **Dorsal fin rays 7-8; anal fin rays 6-9; 24-28 scales in body length; lower jaw slightly projecting** 3

2b **Dorsal fin rays 7-9; anal fin rays 8-10; 29-32 scales in body length; dorsal fin far back, distance from origin to caudal fin is 1/2 distance to snout** ....
.... *Gambusia*

MOSQUITOFISH, *Gambusia affinis* (Baird and Girard). Figs. 475, 476. Light olive, each scale dark edged. Some eastern forms show more pigmentation on sides and under eye. Length about 1 1/2 inches. Apparently 2 subspecies, one from southern Illinois and Indiana south to southern Alabama and Mexico, the other from southern New Jersey into Alabama. Introduced into California and elsewhere, including Hawaii.

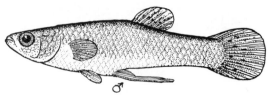

Figure 475.

MANGROVE GUPPY, *Gambusia rhizophorae* Rivas. Brownish with a diffuse lateral band and 4-6, usually 5, longitudinal rows of darks spots. Dorsal fin rays usually 9. Fresh and brackish waters of southern Florida and northwestern Cuba.

Other species of this genus are found in Texas and Mexico. Some have a very limited distribution.

LARGESPRING GAMBUSIA, *Gambusia geiseri* Hubbs and Hubbs. Thin lateral band; dark post anal streak; sides marked with rounded spots. Upper Guadalupe River system and introduced elsewhere in Texas.

CLEAR CREEK GAMBUSIA, *Gambusia heterochir* Hubbs. Indistinct lateral band; dorsal fin rays 7-9. Clear Creek, Menard County, Texas.

PECOS GAMBUSIA, *Gambusia nobilis* (Baird and Girard). Thin lateral band; sides marked with round spots. Western tributaries of the Pecos River, Texas.

BLOTCHED GAMBUSIA, *Gambusia senilis* Girard. Broad lateral band; scales on sides with dark crescents extending below midline. Devils River, Texas, and Sonora and Chihuahua, Mexico.

BIG BEND GAMBUSIA, *Gambusia gagei* Hubbs. Scales on sides above mid-line with dark crescents; broad lateral band. Several springs in Brewster County, Texas.

SAN MARCOS GAMBUSIA, *Gambusia georgei* Hubbs and Peden. Crosshatched by dark pigment and with a diffuse lateral stripe extending from pectoral fin to near base of caudal fin. Dorsal fin rays usually 7. San Marcos River, Texas.

3a Sides marked with lateral band and crossbars; black spot in dorsal and anal fins.....................*Heterandria*
LEAST KILLIFISH, *Heterandria formosa* Agassiz. Fig. 477. Rather olivaceous. One of the smallest fishes in the United States. Length one inch. Swamps and ditches from South Carolina southward and along the Gulf coast to New Orleans.

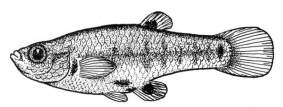
Figure 477.

3b Sides with faint lateral band and no crossbars; fins plain.................
.....................*Poeciliopsis*
GILA TOPMINNOW, *Poeciliopsis occidentalis* (Baird and Girard). Fig. 478. Brownish above dotted with black, silvery below. Length 2 1/2 inches. Gila River system, Arizona into Mexico.

Figure 478.

A number of species of this family from Central America and well known to tropical fish fanciers have been introduced and established in Hawaii and in North America. These include the guppy, *Poecilia reticulata* Peters, known from Blue Point Spring, Clark County and Preston Town Spring, White Pine County, Nevada and Basin Hotspring, Banff National Park, Alberta, Canada. The variable platyfish, *Xiphophorus variatus* (Meek), is found in warm springs in the Missouri River drainage in Montana. In Hawaii, the topminnow, *Limia vittata* (Guichenot), the moonfish, *Xiphophorus maculatus* Gunther and the swordfish, *X. helleri* Heckel have become established. The pike killifish, *Belonesox belizanus* Kner, has been introduced in Florida in the Miami area from Yucatan stock.

SILVERSIDE FAMILY
Atherinidae

The silversides are slender fishes, usually small and with a rather conspicuous silvery band on their sides. Many species occur in the warmer seas and several species occur in fresh water. The dorsal fin is divided into a small spinous portion and a larger soft-rayed portion which are widely separated.

The topsmelt, *Atherinops affinis* Ayres, a common marine form in the Pacific sometimes enters rivers California to Oregon. Several marine species in the Atlantic invade fresh waters and are in the following key.

The freshwater species are pale and rather transparent green in color with a wide silvery longitudinal band on each side. They seldom reach a length of more than 3 inches. They often swim in schools at the surface and may skip short distances out of the water. They spawn in the spring and produce eggs which have a sticky thread enabling them to float until the thread becomes attached to some object.

1a Scales with wrinkled or jagged edges . . . .
. . . . **ROUGH SILVERSIDE,**
Membras martinica **(Valenciennes)**
Greenish silver with narrow lateral band. Resembles the tidewater silverside. Length 4 inches. Marine but enters fresh water along the Gulf of Mexico.

1b Scales with edges smooth 2

2a Scales small, more than 50 rows in body length; snout longer than diameter of eye. Fig. 479 . . . .
. . . . **BROOK SILVERSIDE,**
Labidesthes sicculus **(Cope)**

Figure 479.

Common in lakes and larger streams of the upper Mississippi drainage and the Great Lakes area south to Texas and Florida.

2b Scales large, less than 50 in body length; snout about equal to diameter of eye . . . 3

3a Scales before dorsal fin 14-16 . . . .
. . . . **TIDEWATER SILVERSIDE,**
Menidia beryllina **(Cope)**
Body not as slender as Mississippi silverside. Brackish water but enters river mouths along the coasts of the Atlantic and the Gulf of Mexico.

3b Scales before dorsal fin 15-22 4

4a Scale rows in body length only 39-40. Fig. 480 . . . .
. . . . **MISSISSIPPI SILVERSIDE,**
Menidia audens **Hay**

Figure 480.

Resembles tidewater silverside but body more slender and scales smaller. Brackish water along Gulf of Mexico ascending the Red River in Texas and the Mississippi River drainage as far as Tennessee.

4b Scale rows in body length 44-50 . . . .
. . . . **WACCAMAW SILVERSIDE,**
Menidia extensa **Hubbs and Raney**
Resembles Mississippi silverside except for more scale rows and body not as slender. Lake Waccamaw, North Carolina.

STICKLEBACK FAMILY
Gasterosteidae

The sticklebacks are all small fishes with slender streamlined bodies. They are characterized by a series of free dorsal spines in front of the soft dorsal fin. Some variation occurs in the number of spines. The pelvic fins are reduced to heavy spines. Several species occur in the sea and may enter fresh water. Only one species is restricted to fresh water, another species living equally well in fresh or salt water.

Sticklebacks lack scales. The marine

forms tend to have heavy bony plates on their sides, but the freshwater forms, even the marine forms living in fresh water, have the body naked.

The sticklebacks build elaborate nests. The males construct the nest about the size of a golf ball penetrated by a tunnel used for the eggs. The nest is composed of grasses or fibers cemented together by a secretion of the male. The male cares for the eggs and watches after the newly hatched young.

Sticklebacks are noted for their pugnacious habits attacking fishes many times their size who intrude near their nests. They are predaceous feeding on small or minute aquatic animals.

1a Gill membranes joined to each other forming a fold across the isthmus. Fig. 481. 2

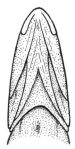

Figure 481.

1b Gill membranes not joined together but joined separately to the isthmus. Fig. 482. 3

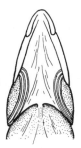

Figure 482.

2a Free dorsal spines usually 9 (8-11). Fig. 483. . . . . NINESPINE STICKLEBACK, *Pungitius pungitius* (Linnaeus)

Figure 483.

Body very slender with long caudal peduncle. Brownish green above, silvery below and irregularly barred. Length 3 inches. Fresh and brackish waters of Northern Hemisphere. In the U.S. it is found in arctic Alaska, the Hudson Bay drainage in Minnesota, the Great Lakes, the headwaters of Mississippi River in Minnesota, several lakes in the Mississippi drainage in Indiana, and as far south as New Jersey on the Atlantic coast.

2b Free dorsal spines usually 5 (4-6). Fig. 484. . . . . BROOK STICKLEBACK, *Culaea inconstans* (Kirtland)

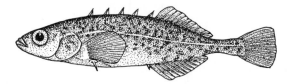

Figure 484.

Body stout but rather streamlined. Brown to black above and light below; mottled on sides and finely speckled below. Length 2 1/2 inches. Western Canada, Montana to Maine and south to Kansas, Illinois, Indiana, and Ohio. Introduced elsewhere, reported from northeastern New Mexico.

3a Free dorsal spines usually 3. Fig. 485 . . . . THREESPINE STICKLEBACK, *Gasterosteus aculeatus* Linnaeus

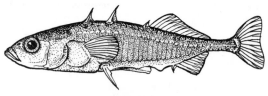

Figure 485.

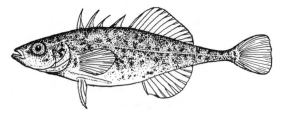

Figure 486.

Body rather stout; brownish green above and light below, profusely speckled with black. Sides may be covered by bony plates or may be naked. Length 4 inches. Fresh water and marine in Northern Hemisphere. In the U.S. it occurs along the Pacific and Atlantic coasts and in the coastal streams northward into the arctics.

**3b Free dorsal spines usually 4. Fig. 486 . . . .
. . . . FOURSPINE STICKLEBACK,
Apeltes quadracus (Mitchill)**

May have 3-4 free dorsal spines; no bony plates in skin but has bony ridge on each side of belly. Body streamlined with very slender caudal peduncle. Brownish green to black above, silvery below and rather mottled. Length 2 1/2 inches. Marine, Atlantic coast from Virginia northward, sometimes entering coastal streams.

TWOSPINE STICKLEBACK, *Gasterosteus wheatlandi* Putnam. Usually with only 2 dorsal spines. In brackish water sometimes entering river mouths, Massachusetts to Newfoundland.

PIPEFISH FAMILY
Syngnathidae

This family contains the pipefishes and sea horses which are found in the warmer seas of the world. Although marine, pipefishes sometimes enter the mouths of rivers and in the southern United States the gulf pipefish, *Syngnathus scovelli* (Evermann and Kendall) (Fig. 487) may be found some distance inland. *Syngnathus leptorhynchus* Girard has been found in California coastal streams. Pipefishes are very long and slender with prehensile tails by which they cling to vegetation. They appear to be jointed as their bodies are covered by

Figure 487.

ring-like plates. Their snout is exceedingly long and bears small jaws at the tip. The males have abdominal pouches in which they carry the eggs placed there by the females. They reach a length of about 5 inches.

SCULPIN FAMILY
Cottidae

The sculpin family is primarily marine, but a number of species are restricted to fresh water. The freshwater forms are characterized by large flat heads and rather slender bodies. The eyes are located on the upper surface of the head and are close together. The soft rays of the dorsal fin are seldom branched and care must be taken to distinguish them from the spines. The pectoral fins are very much enlarged and in some species may have a few branched rays. One of the diagnostic characters is the spines on the edge of the preopercle of most species. These are covered by skin, but can be detected by close scrutiny or by dissection. Sculpins are without scales but may be covered by tiny prickles.

Sculpins are bottom fishes hiding under rocks during the day, either in swift streams or along rocky shores of cold lakes of the mountains or the northern United States. They feed on a wide variety of aquatic organisms including small fishes. Sculpins prepare a nest where the eggs are suspended on the underside of logs or stones and are guarded by the male. Some are reported to be fall spawners. Most species are colored a grayish or olive brown and mottled with dark brown to black. The species are fairly well defined east of the Rockies but are not as clearly defined west of the Rockies, where much variation occurs in different river systems. Identification is sometimes difficult as many characters used in the key are not always constant throughout the range of the species.

1a Gill membranes united (Fig. 488), free from isthmus. . . . .
. . . . FOURHORN SCULPIN, *Myoxocephalus quadricornis* (Linnaeus)

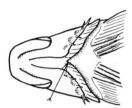

Figure 488.

Pale brown or cream color. Males with very large soft dorsal fin (Fig. 489), smaller in female (Fig. 490). Four sharp preopercular spines. Length to 6 inches. Deep waters of the Great Lakes, arctic streans and coastal waters of North America.

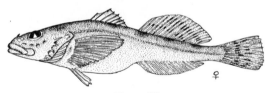

Figure 489.

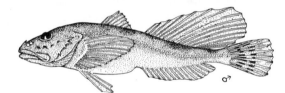

Figure 490.

1b Gill membranes attached to wide part of isthmus (Fig. 491) 2

Figure 491.

2a Upper preopercular spine very large with spinules, antler-like (Fig. 492) . . . .
. . . . **PACIFIC STAGHORN SCULPIN,**
Leptocottus armatus **Girard**

Figure 492.

Grayish olive above and light below; skin smooth. Lateral line complete. Length up to 12 inches. Marine, entering coastal streams, Baja California to Alaska.

2b Upper preopercular spine smaller, without spinules, not antler-like. 3

3a Upper preopercular spine curved upward, crescent-shape (Fig. 493) . . . .
. . . . **SPOONHEAD SCULPIN,**
Cottus ricei **(Nelson)**

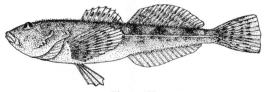

Figure 493.

Body grayish and deeply blotched; heavily covered with prickles. Lateral line complete. Length 3 inches. Shores of Great Lakes and northward into the Yukon Territory.

3b Upper preopercular spine when developed, straight (Fig. 494), not crescent-shaped 4

Figure 494.

4a Pelvic soft rays usually 3 (rarely 4), Fig. 495A 5

4b Pelvic soft rays usually 4 (rarely 3), Fig. 495B. 10

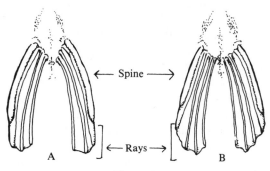

Figure 495.

5a Preopercular spines usually absent or only rudimentary. Fig. 496 . . . .
. . . . **WOOD RIVER SCULPIN,**
Cottus leiopomus **Gilbert and Evermann**

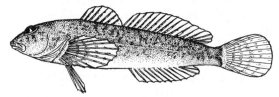

Figure 496.

Grayish olive and mottled; lateral line incomplete. Skin usually smooth, without prickles. Length 3-4 inches. Little Wood River, Shoshone, Idaho.

5b Preopercular spines present, sometimes short and blunt 6

6a Preopercular spines usually 2 or 3; the upper spine rather long . . . . SLENDER SCULPIN, *Cottus tenuis* (Evermann and Meek)

Dark brown with splotches; lateral line incomplete but extending to or beyond last dorsal ray. Some pectoral fin rays are branched. Length 3 inches. Klamath Lake, Oregon.

6b Preopercular spines usually 1 or 2 (rarely 3), the upper one short and blunt and lower spine reduced to a knob. Fig. 497 7

Figure 497.

7a Palatine teeth present . . . . SHOSHONE SCULPIN, *Cottus greenei* (Gilbert and Culver)

Only one developed preopercular spine; pectoral fin rays not branched. Lateral line very incomplete. Snake River, Idaho.

7b Palatine teeth absent 8

8a Anal fin rays usually less than 13. Fig. 498 . . . . SLIMY SCULPIN, *Cottus cognatus* Richardson

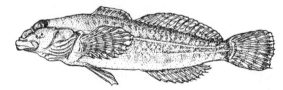

Figure 498.

Olive brown with dark splotches; spinous dorsal fin low with light margin in males. Anal fin rays 10-12. Pectoral fin rays not branched. One large preopercular spine and 1 or 2 small ones below. Lateral line incomplete. Length 3-5 inches. Widely distributed from the Great Lakes region into Alaska. Isolated populations in southeastern Minnesota and northeastern Iowa and in the James and Potomac River drainages, Virginia.

8b Anal fin rays usually 13 or more 9

9a Lateral line incomplete . . . . ROUGH SCULPIN, *Cottus asperrimus* Rutter

Grayish olive to brownish, with 4 or 5 blotches on the sides, belly speckled. Sides covered with prickles. Lateral line not extending beyond posterior end of second dorsal fin, 19-29 pores. Anal fin rays 13-17. Some pectoral fin rays branched. Upper preopercular spine blunt, lower one reduced to a knob (Fig. 497). Pit River, California.

PYGMY SCULPIN, *Cottus pygmaeus* Williams. Grayish black with two ill-defined saddles beneath soft dorsal fin. Lateral line pores 21-22. Two preopercular spines, poorly developed. Total length less than 2 inches. Coldwater Spring, tributary to Coosa River, Alabama.

9b Lateral line complete or nearly complete . . . . MARGINED SCULPIN, *Cottus marginatus* (Bean)

Brownish with blotches. Pectoral fin rays not branched. Anal fin rays 14-16. Large preopercular spine with blunt lobe below. Walla Walla River, Washington.

10a Lateral line pores usually 30 or more . . 11

10b Lateral line pores usually less than 30. . 16

11a Anal fin rays usually 17 or 18. Fig. 499. . . . . PRICKLY SCULPIN, *Cottus asper* Richardson

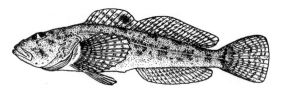

Figure 499.

Grayish olive, heavily mottled; usually heavily prickled. One large preopercular spine with one or two small ones below. Dorsal fins broadly joined; spinous dorsal with 7-10 spines and black spot posteriorly; soft dorsal with 19-23 rays. Lateral line complete, 28 to 43 pores. Palatine teeth present. Length up to 12 inches. Coastal streams from Ventura County, California to Alaska.

11b   **Anal rays usually less than 17 (11-17) . . 12**

12a   **Lateral line incomplete, not reaching past last soft dorsal ray. Fig. 500. . . . . RIFFLE SCULPIN,** *Cottus gulosus* **(Girard)**

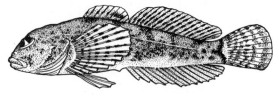

Figure 500.

Grayish olive and mottled; dorsal fin sometimes with a dark spot posteriorly. Anal fin rays 14-16. Lateral line complete or incomplete, 22-36 pores. Length 3-6 inches. Sacramento-San Joaquin system; coastal streams from Morro Bay to Noyo River, California; coastal streams of Oregon and Washington.

12b   **Lateral line complete, reaching past last ray of second dorsal fin 13**

13a   **With palatine teeth 14**

13b   **Without palatine teeth 15**

14a   **Body with 2 dark saddles beneath the second dorsal fin. Fig. 501. . . . . TORRENT SCULPIN,** *Cottus rhotheus* **(Smith)**

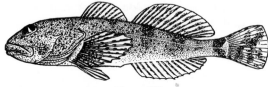

Figure 501.

Dark grayish brown, saddles beneath dorsal extending downward and forward at an angle. Prickles usually present on sides and back. Dorsal fins slightly separated. Preopercular spine short and heavy with 2 small spines below. Length 3-4 inches. Lower Columbia River and Puget Sound drainages.

14b   **Body with 4 prominent dark saddles. Fig. 502. . . . . BANDED SCULPIN,** *Cottus carolinae* **(Gill)**

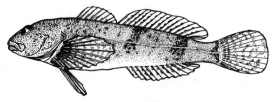

Figure 502.

Light brown, one saddle anterior to base of spinous dorsal, one at anterior base and another at posterior base of soft dorsal, fourth saddle just anterior to base of caudal fin. Lateral line pores 30-34. One large preopercular spine with small spines below. Length to 4 inches. Upper Tennessee River drainage to the Ozarks.

15a Soft rays in dorsal fin usually 17 or more; pelvic fins long, reaching vent. Fig. 503 .. . . COAST RANGE SCULPIN, *Cottus aleuticus* Gilbert

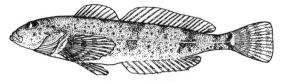

Figure 503.

Grayish olive with mottling on back, 2 or 3 vertical bands beneath second dorsal fin, belly white. Second dorsal with 17-20 rays. Lateral line pores 34-44. Preopercular spine rather short with no spines below it. Patch of prickles posterior to pectoral fin. Length to 5 inches. Pacific coastal streams from San Luis Obispo County, California to the Aleutian Islands.

15b Soft rays in dorsal fin usually less than 17; pelvic fins shorter, not reaching vent. Fig. 504 . . . . PIUTE SCULPIN, *Cottus beldingi* Eigenmann and Eigenmann

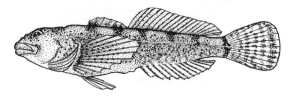

Figure 504.

Grayish brown and mottled with dark saddles; skin smooth, no prickling. Pectoral fin rays unbranched. Anal fin rays 11-13. With a single preopercular spine. Length 3-4 inches. Lahontan System in California and Nevada; Lake Tahoe and its tributaries; Columbia River, Oregon and Washington; Bear River, Utah and Idaho. Upper Colorado River in Colorado.

PIT SCULPIN, *Cottus pitensis* Bailey and Bond. Similar to the riffle sculpin. Color variable, with 5-6 faint saddles on back and a dark band encircling caudal peduncle. Lateral line pores 33-37. Two preopercular spines. Pit River System, Goose Lake, Oregon to Sacramento River, Shasta County, California.

16a Lateral line pores 22 or less . . . . MARBLED SCULPIN, *Cottus klamathensis* Gilbert

Marbled olive brown with barred fins. Dorsal fins more or less confluent. Some pectoral fin rays branched. Lateral line pores 14-22. One well-developed preopercular spine, a smaller one sometimes present below. Klamath and Lost River systems, tributaries of Pit River, California.

POTOMAC SCULPIN, *Cottus girardi* Robins. Pale brownish on back and sides, two narrow bands across back and under second dorsal fin, belly pale. Lateral line incomplete, 17-25 pores. Three preopercular spines. Length up to 5 inches. Tributaries of the Potomac River System, Pennsylvania, Maryland, Virginia and West Virginia.

16b Lateral line pores more than 22 17

17a Rays in second dorsal fin 20-21. Fig. 505 . . . . KLAMATH LAKE SCULPIN, *Cottus princeps* Gilbert

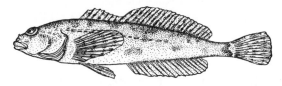

Figure 505.

Preopercular spine reduced to a small knob or absent; dorsal fins confluent; dorsal spines usually 7; some pectoral fin rays branched. Pelvic fins long, reaching to vent. Lateral line incomplete. Upper Klamath Lake, Oregon.

17b Rays in the second dorsal fin 14-19. . . . 18

18a Palatine teeth absent . . . .
. . . . **RETICULATE SCULPIN,**
Cottus perplexus **Gilbert and Evermann**
Grayish brown with faint vermiculation color pattern and dark spots. Lateral line complete or incomplete, 22-32 pores. One to 4 preopercular spines. Dorsal fins usually joined; dorsal fin rays 18-20. Pectoral fin rays 13-16. Northern Oregon and Washington.

BLACK SCULPIN, *Cottus baileyi* Robins. Blackish to grayish with lighter belly with pigment. Dorsal spines 7-8; dorsal rays 16-17 (15-18). Lateral line pores usually less than 30 (22-33). Upper preopercular spine usually developed, lower 2 reduced. Length less than 3 1/2 inches. Headwaters of Holston River, Tennessee and Virginia.

18b Palatine teeth present (concealed narrow row in *Cottus confusus*) **19**

19a Prickles numerous, present on back and sides, belly may be naked **20**

19b Prickles present or absent, when present usually axillary in a patch near pectoral fin **21**

20a Prickles widely distributed over body including breast and belly . . . .
. . . . **UTAH LAKE SCULPIN,**
Cottus echinatus **Bailey and Bond**
Brownish without saddles or blotches. Dorsal spines 7, rarely 8. Pectoral fin rays 16-18. Lateral line pores 26-29. Restricted to Utah Lake, Utah.

20b Prickles on back and sides and belly naked . . . .
. . . . **BEAR LAKE SCULPIN,**
Cottus extensus **Bailey and Bond**
Brownish with a few blotches. Dorsal spines 7 or 8. Pectoral fin rays 15-17. Lateral line pores 22-31. Three preopercular spines and occa-

sionally a fourth represented by a knob. Bear Lake, Utah and Idaho.

21a Pectoral fin rays usually 14-16; preopercular spines 3. Fig. 506 . . . .
. . . . **MOTTLED SCULPIN,**
Cottus bairdi **Girard**

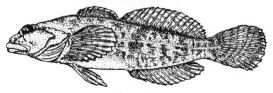

Figure 506.

Dark brown to grayish olive; spiny dorsal fin with a black band forming a heavy spot posteriorly, fin edged with pink in breeding males. One large preopercular spine with 2 or 3 small spines below. Lateral line incomplete, pores 19-26. Length 4-6 inches. Widespread east of the Rockies from northwestern Canada to the Ozarks and the southern Appalachians. Represented in the west by two sub-species in the upper Colorado River drainage and the upper Columbia River.

21b Pectoral fin rays 13 or 14; preopercular spines 2, sometimes 3. Fig. 500 . . . .
. . . . **RIFFLE SCULPIN,**
Cottus gulosus **Girard**
Mottled grayish olive; first dorsal fin with dark blotch on posterior. Anal fin rays 14-16. Lateral line complete or incomplete, 22-36 pores. Length 3-6 inches. Sacramento-San Joaquin System, coastal streams from Morro Bay to Noyo River, California; coastal streams of Oregon and Washington.

SHORTHEAD SCULPIN, *Cottus con-fusus* Bailey and Bond. Grayish brown and slightly mottled. Pectoral rays 13-14; anal fin rays 12-14. Lateral line pores 22-33. Large preopercular spine with a small spine below. Length to 6 inches. Puget Sound and Columbia River basins, Flathead River, Montana.

TEMPERATE BASS FAMILY
Percichthyidae

Formerly treated as part of the sea bass family, Serranidae. This family closely resembles the Sunfish Family to which the black basses belong, and is differentiated externally by several rather obscure characters. The freshwater species do not build nests, but spawn at random in the spring and give no care to the eggs or young. The dorsal fin is often completely divided, and the dorsal spinous portion is higher than in the sunfish family. Also, they have several skeletal differences and possess well-developed pseudobranchiae.

1a Dorsal fins separate; soft anal rays 11 or 12; lower jaw projecting 2

1b Dorsal fins joined; soft anal rays 8 to 10; jaws about equal 3

2a Depth of body more than 1/3 standard length; hyoid teeth usually in a single patch. Fig. 507 . . . . WHITE BASS, *Morone chrysops* (Rafinesque)

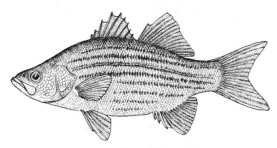

Figure 507.

Silvery color, sides with about 7 longitudinal stripes which may be broken. Reaches length of 18 inches. Minnesota east through lower Great Lakes and St. Lawrence drainage and south to northern Alabama and to Texas. Introduced elsewhere.

2b Depth of body less than 1/3 standard length; hyoid teeth in two patches. Fig. 508 . . . . STRIPED BASS, *Morone saxatilis* (Walbaum)

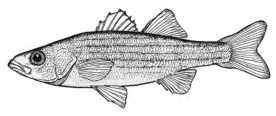

Figure 508.

Back olivaceous; sides silvery or brassy, marked with about 7 dark longitudinal lateral stripes. Reaches large size in the sea, weighing over 100 pounds. Anadromous, entering streams of Atlantic coast. Introduced and common on Pacific coast.

3a Sides with about 7 longitudinal black stripes, those below lateral line broken under dorsal fin; longest dorsal spine more than 1/2 the length of the head. Fig. 509 . . . . YELLOW BASS, *Morone mississippiensis* Jordan and Evermann

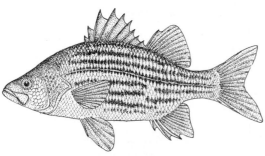

Figure 509.

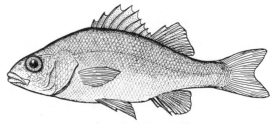

.... **WHITE PERCH,**
Morone americana (Gmelin)

Figure 510.

Brassy yellow. Length 18 inches. Southern Minnesota to Ohio and southward in Mississippi Valley.

3b   Sides nearly plain, may have faint light streaks; longest dorsal spine about 1/2 length of head. Fig. 510.............

Back olivaceous; sides more or less silvery. Length 14 inches. Marine, common along the Atlantic coast southward to South Carolina, frequently entering rivers and connected bodies of waters. May become landlocked.

SUNFISH FAMILY
Centrarchidae

The sunfish family contains not only the sunfishes, but also the crappies and black basses. Most of these are either game or pan fishes. The members of this family were originally not found west of the Rocky Mountains except for the Sacramento perch in the central valley of California. Many species of this family have been introduced all over the world.

The sunfishes, crappies and black basses prefer the warmer lakes and streams from southern Canada to the Gulf. All except the Sacramento perch are nesting fishes. The males scoop out a depression where one or more females deposit eggs. The males guard the eggs and the newly hatched young.

The members of this family closely resemble those of the perch family and the bass family. They differ in that the spinous and soft portions of the dorsal fin are united and confluent,

except in the largemouth bass where a deep notch almost separates the two parts.

1a   Body elongated; depth goes about 3 times in standard length 2

1b   Body short and deep; depth goes less than 3 times in standard length 7

2a   Lateral line absent; dorsal fin not deeply notched; caudal fin rounded; scale rows less than 50, PYGMY SUNFISHES, *Elassoma*......................3**

2b   Lateral line developed and present; dorsal fin almost divided; caudal fin notched; lateral line scales more than 58........4

3a Scale rows in body length 35 to 45; Fig. 511. . . . .
. . . . **BANDED PYGMY SUNFISH,**
Elassoma zonatum **Jordan**

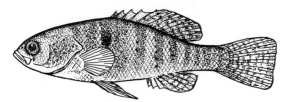

Figure 511.

Olivaceous with about 11 vertical bars on sides; black spot size of eye on side under front of dorsal fin. Length 1 1/2 inches. Southern Illinois to Texas and Florida.

3b Scale rows in body length 27 to 30; Fig. 512. . . . .
. . . **EVERGLADES PYGMY SUNFISH,**
Elassoma evergladei **Jordan**

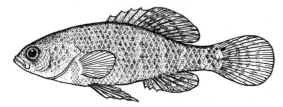

Figure 512.

Olivaceous with variable markings. Length 1 1/2 inches. Swamps of southern Georgia and Florida.

OKEFENOKEE PYGMY SUNFISH, *Elassoma okefenokee* Böhlke. Brownish with bright blue, very similar to Everglades pygmy sunfish but it tends to have more dorsal rays, no scales on top of head and only 1-2 rows of scales on cheek. Okefenokee swamp, Georgia and in northern Florida.

4a Upper jaw extends past posterior margin of orbit; spinous dorsal fin separated from soft dorsal fin by a deep notch ex-

tending almost to base of fin. Fig. 513 . . .
. . . . **LARGEMOUTH BASS,**
Micropterus salmoides **(Lacépède)**

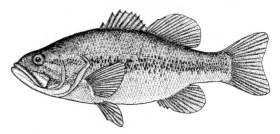

Figure 513.

Dark green above, sides and belly silvery; dark lateral band present, breaks up in old individuals. Length up to 20 inches or more. Southern Canada through Great Lakes drainage and south into Mexico; Virginia to Florida on Atlantic coast. Widely introduced elsewhere including Hawaii.

4b Upper jaw does not extend behind posterior margin of orbit; spinous dorsal fin incompletely separated by shallow notch from soft dorsal fin 5

5a Soft dorsal rays usually 13-15. Fig. 514 . .
. . . . **SMALLMOUTH BASS,**
Micropterus dolomieui **Lacépède**

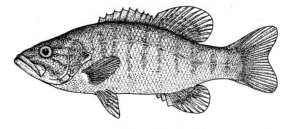

Figure 514.

Greenish above, dusky silver below; each scale sometimes with a brassy spot. Length up to 18 inches. Minnesota to Quebec and south to Arkansas and northern Alabama. Widely introduced elsewhere including Hawaii.

5b Soft dorsal rays usually 12 (11-13) 6

6a Juvenile body with lateral band; adults usually with broken row of lateral blotches, more or less connected. Fig. 515 . . . . **SPOTTED BASS,** *Micropterus punctulatus* **(Rafinesque)**

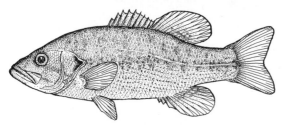

Figure 515.

Greenish above, silvery below and somewhat mottled. Length up to 17 inches. Southern Illinois, Missouri, and Ohio southward to eastern Texas and Gulf. Introduced into Central Valley reservoirs in California.

SUWANNEE BASS, *Micropterus notius* Bailey and Hubbs. Body with broken lateral row of blotches as in spotted bass. Basicaudal spot present, otherwise resembles smallmouth bass. Ichtucknee Springs, Columbia County, Florida.

6b Juvenile body with vertical bars on side but which may be broken into row of blotches in adults. . . . . **REDEYE BASS,** *Micropterus coosae* **Hubbs and Bailey**
Body with vertical bars on sides, each bar with a light center; notch separating dorsal fins very slight. Basicaudal spot not very distinct. Upland streams of Georgia and Alabama south into western Florida.

GUADALUPE BASS, *Micropterus treculi* (Valliant and Boucourt). Sides with vertical bars which become broken into one or more rows of blotches in adults; basicaudal spot prominent in young but faint in adults. Central Texas.

7a Base of dorsal fin only slightly longer than anal fin 8

7b Base of dorsal fin much longer than base of anal fin 10

8a Dorsal spines more than 10; anal spines usually 7 or 8. Fig. 516. . . . . **FLIER,** *Centrarchus* *macropterus* **(Lacépède)**

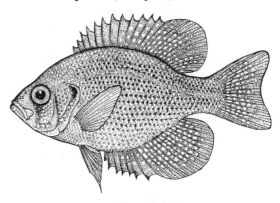

Figure 516.

Body quite deep; greenish, each scale with a brown spot giving appearance of numerous rows of dots on sides. Length 6-7 inches. Virginia to Florida, and southern Illinois southward in Mississippi valley.

8b Dorsal spines less than 10; anal spines usually 6 9

9a Distance from eye to front of dorsal fin base about equal to base of dorsal fin. Fig. 517. . . . . **BLACK CRAPPIE,** *Pomoxis nigromaculatus* **(Lesueur)**

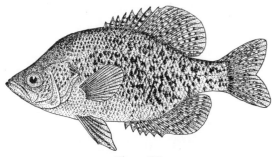

Figure 517.

Silvery, mottled with dark green or black; vertical fins spotted. Length up to 12 inches. Upper Mississippi Valley and Great Lakes, southward to Florida and Texas. Widely introduced elsewhere.

9b **Distance from eye to front of dorsal fin base greater than length of base of dorsal fin. Fig. 518.** . . . .
. . . . **WHITE CRAPPIE,**
Pomoxis annularis **Rafinesque**

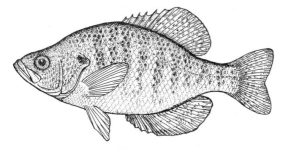

Figure 518.

Silvery white, mottled with dark green or black and with vertical bars on sides. Length up to 12 inches. Southern Minnesota and Great Lakes region south to Texas and western Florida. Widely introduced.

10a **Caudal fin more or less rounded** 11

10b **Caudal fin more or less forked** 14

11a **Scales cycloid; mouth large, maxillary extending behind middle of eye. Fig. 519.** . . . . **MUD SUNFISH,**
Acantharchus pomotis **(Baird)**

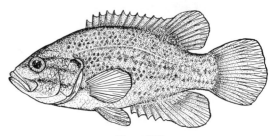

Figure 519.

Dark greenish with 5 rather indistinct dark longitudinal bands along sides. Length up to 6 inches. Lowland streams, New York to northern Florida.

11b **Scales ctenoid; mouth small, maxillary not extending behind middle of eye** . . . 12

12a **Dorsal spines 10; front of dorsal fin black. Fig. 520.** . . . .
. . . . **BLACKBANDED SUNFISH,**
Enneacanthus chaetodon **(Baird)**

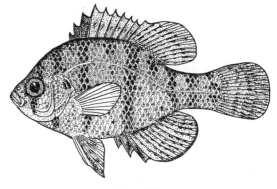

Figure 520.

Straw color, sides strongly banded with transverse bars, the bar behind the shoulder extending over the front of the dorsal fin. Length about 4 inches. Lowland streams from New Jersey to Florida.

12b Usually 9 dorsal spines; no black streak at front of dorsal fin **13**

13a Sides with 5 to 8 distinct crossbars; numerous pale or light blue spots on spiny part of dorsal fin. Fig. 521. . . . .
. . . . **BANDED SUNFISH,**
Enneacanthus obesus **(Girard)**

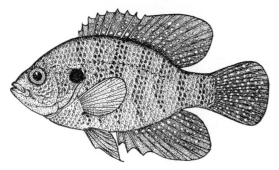

Figure 521.

Opercular spot large as eye. Olivaceous, with purplish or golden spots on body and fins. Length about 3 inches. Coastal lowlands, Massachusetts to Florida.

13b Sides with indistinct crossbars; none or very few pale or light blue spots on spiny part of dorsal fin. Fig. 522. . . . .
. . . . **BLUESPOTTED SUNFISH,**
Enneacanthus gloriosus **(Holbrook)**

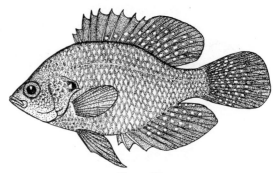

Figure 522.

Opercular spot smaller than eye. Dark olivaceous with bright blue spots on sides and fins; pearly spot in front of opercular spot. Length about 2 3/4 inches. Coastal lowlands, New Jersey to Florida.

14a Mouth large, maxillary extending behind middle of eye **15**

14b Mouth small, maxillary not extending behind middle of eye. The SUNFISHES, *Lepomis* **17**
The various species of this genus hybridize readily and the hybrids will not fit this key.

15a Anal spines 3. Fig. 523. . . . .
. . . . **WARMOUTH,**
Chaenobryttus gulosus **(Cuvier)**

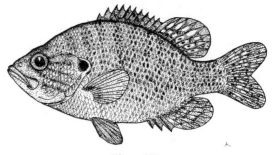

Figure 523.

Brassy-dark olive green, sometimes mottled; each scale with a dark spot. Length up to 10 inches. Southern Minnesota, Great Lakes region south to Texas and Florida, Colorado River and Sacramento-San Joaquin system in California.

15b Anal spines 5 to 8. 16

16a Longest anal spine about half the length of the spinous portion of the anal fin; gill rakers 10. Fig. 524. . . . . ROCKBASS, *Ambloplites rupestris* **(Rafinesque)**

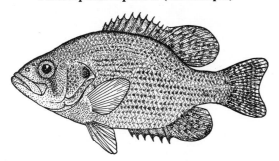

Figure 524.

Brassy-olivaceous and somewhat mottled; each scale with a dark spot forming numerous rows of dots. Length up to 12 inches. Vermont to Saskatchewan and south to the Gulf of Mexico. Several subspecies in the south.

ROANOKE BASS, *Ambloplites cavifrons* Cope. Cheeks naked or incompletely scaled; 10-12 rows of scales above lateral line instead of 7-9. Roanoke and perhaps James River, Virginia.

16b Longest anal spine about equal to length of spinous part of anal fin; gill rakers 20. Fig. 525. . . . . SACRAMENTO PERCH, *Archoplites interruptus* **(Girard)**

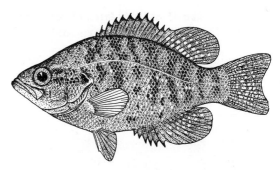

Figure 525.

Dark back, silvery below with about 7 vertical bars on each side; bars are somewhat interrupted and irregular. Length up to 20 inches. Sacramento and San Joaquin River drainages, California. Introduced elsewhere in California and Nevada.

17a Pectoral fin short and rounded (pointed in redbreast sunfish), not reaching behind front of anal fin (except in immature individuals). 18

17b Pectoral fin long and pointed, reaching behind front of anal fin (except in immature individuals) 23

18a Gill rakers long, length about 6 times width of base. Fig. 526. 19

Figure 526.

18b Gill rakers short, length not more than 3 times width of base 20

19a More than 45 scales in lateral line; lateral line complete; body oblong and stout. Fig. 527 . . . . **GREEN SUNFISH,** *Lepomis cyanellus* **Rafinesque**

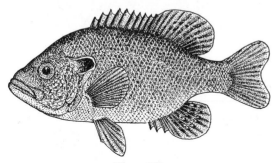

Figure 527.

Brassy-green; opercular lobe dark with light bronze margin. Length about 6 inches. Minnesota and Great Lakes region south to Mexico, not east of the Alleghenies.

19b Less than 40 lateral line scales; lateral line interrupted, many scales not pored; body rounded. Fig. 528 . . . . **BANTAM SUNFISH,** *Lepomis symmetricus* **Forbes**

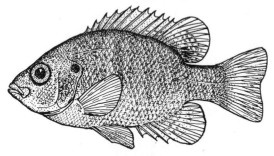

Figure 528.

Sides more or less barred; young have a dark ocellated spot at posterior base of dorsal fin; each scale marked with brown giving appearance of numerous rows of dots. Length about 3 inches. Southern Illinois to Louisiana and Texas.

20a Opercular lobe no longer than wide . . . 21

20b Opercular lobe longer than wide except in immature individuals 22

21a Opercular lobe soft, stiff only at base; length of gill rakers not more than 2 × width of base (Fig. 529); each scale pigmented at base. Fig. 530 . . . . **DOLLAR SUNFISH,** *Lepomis marginatus* **(Holbrook)**

Figure 529.

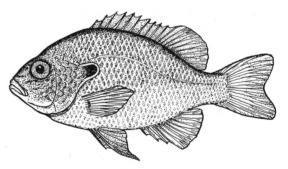

Figure 530.

Olivaceous with orange on cheeks and belly, also numerous blue streaks; opercular lobe long and with pale greenish margin. Length 6-7 inches. Oklahoma to South Carolina and Florida.

21b Opercular lobe stiff most of its length; length of gill rakers about 3 × width of base (Fig. 531); each scale pigmented at base. Fig. 532 . . . . **SPOTTED SUNFISH,** *Lepomis punctatus* **(Valenciennes)**

Figure 531.

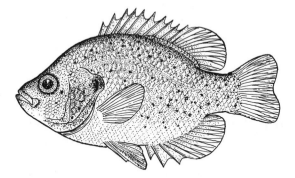

Figure 532.

Olivaceous with numerous brown or black specks scattered over sides of body, each scale may have a tiny spot; opercular spot plain; opercular lobe rather short. Length about 6 inches. South Carolina to Florida. The subspecies *L. p. miniatus* Jordan has fewer specks and is common in Mississippi valley from southern Illinois to Texas.

22a Gill rakers short and knobby (Fig. 533), rather soft; opercular lobe as wide or wider than eye; lateral line scales 36-45. Fig. 534. . . . . LONGEAR SUNFISH, *Lepomis megalotis* (Rafinesque)

Figure 533.

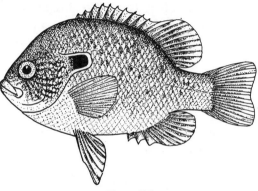

Figure 534.

Brightly colored with orange spots and blue streaks; opercular lobe usually very long and may or may not have bluish colored margin. Length about 8 inches. Quetico region, Ontario, Iowa to southern Quebec and south to south Carolina and into Mexico.

22b Gill rakers short but hard or stiff (Fig. 535); opercular lobe not as wide as eye; lateral line scales 43-50. Fig. 536. . . . . REDBREAST SUNFISH, *Lepomis auritus* (Linnaeus)

Figure 535.

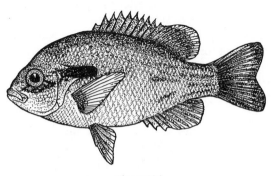

Figure 536.

23a Gill rakers long (Fig. 537), length more than 2 × width of base 24

Figure 537.

23b Gill rakers short (See Fig. 542.), length less than 2 × width of base 25

24a Lateral line scales more than 40; anal soft rays usually 10-12. Fig. 538 . . . . **BLUEGILL,** *Lepomis macrochirus* **Rafinesque**

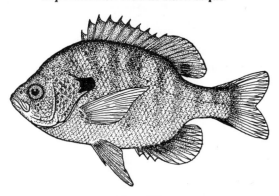

Figure 538.

Olive green with some blue and orange on body; dark spot at posterior base of dorsal fin; vertical bars on sides; opercular lobe solid black; gill rakers long. (See Fig. 537.) Reaches length of over 10 inches. Widespread from Minnesota to Lake Champlain and south to Florida and Texas. Widely introduced elsewhere, including Hawaii.

24b Lateral line scales less than 40; anal soft rays 7-10. Fig. 539 . . . . **ORANGESPOTTED SUNFISH,** *Lepomis humilis* **(Girard)**

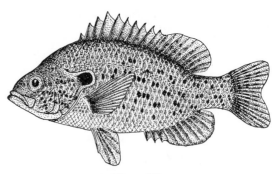

Figure 539.

Body brightly spotted with orange, opercular lobe with broad red or orange margin; pectoral fin about as long as head; gill rakers long (Fig. 540). Length about 4 inches. North Dakota to western Ohio and south to Texas and northern Alabama.

Figure 540.

25a Opercular lobe with a spot of orange or red in the lower part; pectoral fin 3 or more times in standard length. Fig. 541 . . . . **PUMPKINSEED,** *Lepomis gibbosus* **(Linnaeus)**

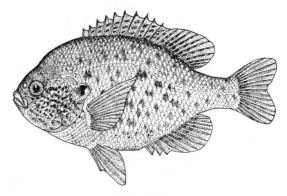

Figure 541.

Very brightly colored with orange and some blue; gill rakers short (Fig. 542). Length over 8 inches. Manitoba and North Dakota to New Brunswick and south to South Carolina, Ohio, and Iowa. Present in the Klamath River drainage, California.

Figure 542.

25b **Opercular lobe with broad red or orange margin below and behind; pectoral fins 3 or less times in standard length. Fig. 543 . . . . REDEAR SUNFISH,** *Lepomis microlophus* **(Gunther)**

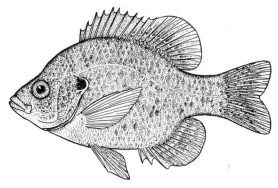

Figure 543.

Olivaceous with yellow or orange breast: gill rakers short (Fig. 544). Length of 10 inches. Missouri to southern Indiana and south to Florida and Texas. Present throughout California.

Figure 544.

PERCH FAMILY
Percidae

This is an important group of freshwater fishes characterized by a dorsal fin which is completely divided into a spiny and a separate soft-rayed portion. The anal fin bears one or two spines. The family consists of three divisions or subfamilies: (1) Percinae, the perch; (2) Luciopercinae, the walleye and sauger; (3) the Etheostominae, the darters.

All members of this family spawn in the spring in a variety of ways. The perch string their eggs in gelatinous strings over the vegeta-

tion. The pikeperches deposit their eggs at random in shallow water. Some darters (logperch, Iowa darter, and least darter) do likewise, while others (rainbow darter) cover their eggs with gravel or sand. Some, such as the johnny darter and the fantail darter, place their eggs on the underside of objects where they are cared for by the males. In many of the darters, the males assume brilliant colors in the spawning season.

The numerous species of the small darters are confined to east of the Rockies. Their taxonomy is difficult and they constitute a "happy hunting ground" for taxonomists. Some species exist which have never been named and some named species resemble each other so closely that they will probably be proven to be sub-species or the same.

All members of this family are predaceous, the larger species being highly piscivorous, and the smaller darters preying on minute insects and crustacea.

1a Branchiostegal rays 7 or 8; pectoral fins normal size; upper jaw extending to or behind middle of eye; pseudobranchiae well developed; size usually large, adults more than 5 inches in length. 2

1b Branchiostegal rays 5 or 6; pectoral fins larger than usual; upper jaw not extending to middle of eye; pseudobranchiae absent or poorly developed; small fishes, adults seldom exceeding 6 inches in length, usually much smaller 3

2a Body with prominent crossbars (may be faint in young); pelvic fins close together, space between less than width of base of either fin; without oversize or fang-like teeth (canines) . . . . *Perca*
YELLOW PERCH, *Perca flavescens* (Mitchill). Fig. 545. Yellowish with 6-7 dark vertical bars. May reach a length of 15 inches. Probably the same as the old world species, *Perca fluviatilis* Linnaeus.

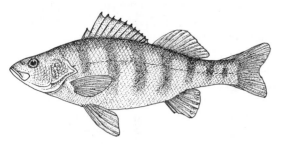

Figure 545.

Widely distributed from northern Kansas to Ohio and South Carolina and northward into Canada. Widely introduced elsewhere.

2b Body without strong crossbars; pelvic fins widely separated by space equal to width of base of either fin; many elongated or fang-like (canine) teeth . . . .
. . . . *Stizostedion*
WALLEYE, *Stizostedion vitreum* (Mitchill). Fig. 546. Lower lobe of caudal fin whitish; dark spot at posterior end of spinous dorsal fin; no black spot on basal part of pectoral fin; 3 or 4 pyloric caeca (finger-like structures where intestine leaves stomach). Reaches length of 36 inches. Tennessee River drainage northward into southern Canada and northwestward to Great Slave Lake and British Columbia.

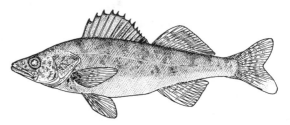

Figure 546.

SAUGER, *Stizostedion canadense* (Smith). Fig. 547. Lower lobe of caudal fin not whitish; no dark spot at posterior base of spinous dorsal fin; black spot on basal portion of pectoral fin; 5-6 pyloric caeca. Reaches length of about 15 inches. Alberta to New Brunswick and south to Oklahoma northern Louisiana and Tennessee River drainage. Introduced elsewhere.

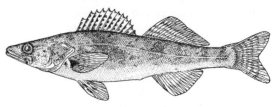

Figure 547.

3a Depth of body goes 7 or more times in length and anal fin has only one spine; scales on trunk tend to be limited to middle of sides; may have several, although usually only one row of scales below lateral line anterior to anus except in crystal darter which has only the belly naked . . . .*Ammocrypta*

Slender darters which are rather pellucid.

CRYSTAL DARTER, *Ammocrypta asprella* (Jordan). Fig. 548. Lateral line scales 89-100; anal rays 12-14. Length 4-5 inches. Southern Minnesota to southern Ohio and south to Alabama and Oklahoma. Other members of this genus have less than 80 lateral line scales and less than 10 anal rays. They seldom reach more than 5 inches in length.

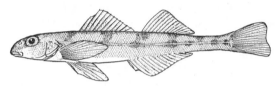

Figure 548.

EASTERN SAND DARTER, *Ammocrypta pellucida* (Putnam). Fig. 549. Cheeks and opercles scaly; peduncle entirely scaled; anterior number of scale rows irregular and mostly confined to above lateral line. Row of dots on sides well defined and not connected. Great Lakes and Ohio River drainages.

Figure 549.

WESTERN SAND DARTER, *Ammocrypta clara* Jordan and Meek. Fig. 550. Checks and opercles more or less scaly; peduncle entirely scaled; scales on trunk confined anteriorly to several rows above lateral line. Southern Minnesota to Indiana and southward to eastern Texas.

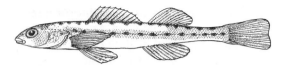

Figure 550.

SCALY SAND DARTER, *Ammocrypta vivax* Hay. Fig. 551. Cheeks and opercles scaly; peduncle entirely scaled; anteriorly scales cover most of sides above lateral line, covering back in front of dorsal fin. Row of spots on sides diffuse and not connected. Oklahoma and Mississippi to Texas.

Figure 551.

NAKED SAND DARTER, *Ammocrypta beani* Jordan. Fig. 552. Cheeks and opercles without scales; scales on trunk and peduncle limited to several rows on and above lateral line. No row of spots on sides; dark blotch in anterior part of spinous dorsal fin. Coastal streams, Mississippi to Florida.

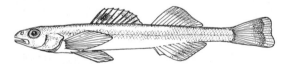

Figure 552.

3b Depth of body usually goes less than 7 times in body length and anal fin has one or two spines, or if depth goes over 7 times in body length, anal fin has 2 spines; scales on trunk covering most of body (nape may be naked); sides well scaled below lateral line, although belly may be more or less naked in some species 4

4a Belly may be naked except for a row of enlarged or modified scales on the mid-ventral line (Fig. 553) or belly may be

more or less scaled with one or more modified scales between the pelvic fins (Fig. 554); area of anal fin usually as large as area of soft dorsal fin; pelvic fins in most species widely separated, space between them usually at least 3/4 as wide as base of either fin . . . .
. . . . (p. 170) *Percina*

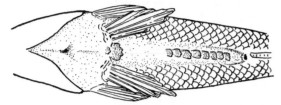

Figure 553.

Figure 554.

Many species in this genus which includes those formerly in *Hadropterus* and several other genera.

4b Belly usually scaled, but if naked no mid-ventral row of modified scales present or no modified scales present between pelvic fins (Fig. 555); space between pelvic usually less than 3/4 as wide as the base of either fin (except in several species); area of anal fin usually less than area of soft dorsal fin . . . .
. . . . (p. 175) *Etheostoma*

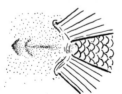

Figure 555.

GENUS *PERCINA*

1a Snout extends before upper lip; sides marked with numerous crossbars or with spots. Fig. 556. . . . .
. . . . . . LOGPERCH, *Percina caprodes* (Rafinesque)

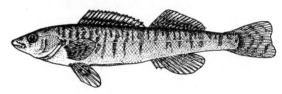

Figure 556.

Yellowish with sides marked by many vertical bars. Belly naked with a mid-ventral row of modified scales. Reaches a length of 6 inches. Several sub-species, Minnesota to Vermont and south to Mississippi, Texas, and western Florida.

BIGSCALE LOGPERCH, *Percina macrolepida* Stevenson. This species is very similar to the logperch, but differs in having many narrow lateral bars of the same length. Texas, Oklahoma and northeastern Mexico. Introduced into California, where it is found in the Sacramento-San Joaquin system.

BLOTCHSIDE LOGPERCH, *Percina burtoni* Fowler. Another species similar to *P. caprodes*. Brownish in color with large blackish brown blotches, belly white. Holston River, Virginia and Tennessee, Swannanoa River, North Carolina.

ROANOKE LOGPERCH, *Percina rex* (Jordan and Evermann). Similar to logperch except sides are marked with rows of spots. Headwaters of Roanoke River, Virginia.

1b Snout does not extend before upper lip; sides usually marked with longitudinal row of large blotches which tend to be confluent along the lateral line. 2

2a Premaxillaries usually protractile, separated from snout by a groove, in some a slight fleshy bridge or frenum may be hidden in the groove 3

2b Premaxillaries not protractile, connected with snout by a smooth fleshy bridge or frenum 5

3a Spiny dorsal fin separated from soft dorsal fin by a wide space. Fig. 557 . . . . CHANNEL DARTER, *Percina copelandi* (Jordan)

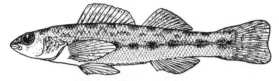

Figure 557.

Some specimens may show a frenum or fleshy bridge connecting premaxillaries to snout. Brownish with row of small blotches on sides and a tiny black spot at base of caudal fin. Length 2 1/2-3 inches. Eastern Michigan to upper St. Lawrence drainage and south on western side of Appalachians to northern Alabama and to Oklahoma.

3b Spiny dorsal fin separated by only a deep notch from soft dorsal fin 4

4a Cheeks scaled. Fig. 558 . . . . RIVER DARTER, *Percina shumardi* (Girard)

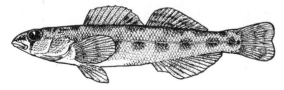

Figure 558.

Rather dark with 8-10 lateral blotches. Length 3 inches. Southern Manitoba and western Ontario to Ohio and south to central Texas and northern Alabama.

SNAIL DARTER, *Percina tanasi* Etnier, Fig. 559. Brownish to brownish gray in color with some green; four prominent brownish saddles present; belly white. Cheeks typically with embedded scales posterior to eye. Little Tennessee River, Tennessee.

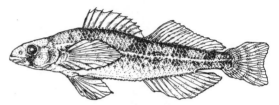

Figure 559.

4b Cheeks naked. Fig. 560 . . . . STARGAZING DARTER, *Percina uranidea* (Jordan and Gilbert)

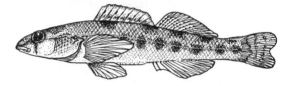

Figure 560.

Greenish olive with 9-10 lateral spots and a small spot at base of the caudal fin. Length 2 1/2 inches. White and Saline rivers, Arkansas.

Percina ouachitae (Jordan and Gilbert). Similar to the stargazing darter. Five well-developed dorsal saddles; 7 or 8 oval spots on sides, often confluent. Small intense caudal spot present. Lower Mississippi Valley from Tennessee to the Gulf Coast drainage.

5a Belly scaled and with no mid-ventral row of modified scales, but with scales like those of the sides; a more or less modified scale present between the pelvic fins. (See Fig. 554.) 6

5b Belly more or less naked and with a mid-ventral row of enlarged or modified scales or with a modified scale between the pelvic fins. (See Fig. 555.) 8

6a Gill membranes scarcely connected (Fig. 561). 7

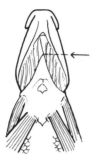

Figure 561.

6b Gill membranes connected. (Fig. 562) . . . . OLIVE DARTER, *Percina squamata* (Gilbert and Swain)

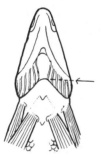

Figure 562.

Yellowish olive with 10 dusky blotches along each side and with a diffuse blotch with a small balck dot behind it at the base of the caudal fin. Conspicuous black humeral spot present. Length 5 inches. Headwaters of the Tennessee River.

7a Snout short and blunt (Fig. 563) . . . . YELLOW DARTER, *Percina aurantiaca* (Cope)

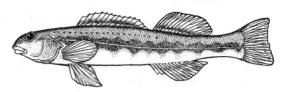

Figure 563.

Olive above and yellowish below; a row of confluent blotches with row of dots above along each side; a black spot at base of caudal fin. Males with much orange. Length 5-6 inches. Headwaters of Tennessee River.

7b Snout long and sharp. Fig. 564 . . . . BLUESTRIPE DARTER, *Percina cymatotaenia* (Gilbert and Meek)

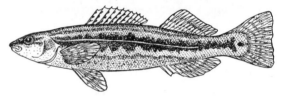

Figure 564.

Yellowish green above and heavily speckled, and with a rather dark longitudinal band on each side and a small black spot at base of caudal fin. Length 4-5 inches. Western Kentucky, southern Missouri and Arkansas.

8a Gill membranes not closely connected. Fig. 561. 9

8b Gill membranes closely connected. Fig. 562 12

9a Cheeks entirely naked 10

9b Cheeks scaled more or less 11

10a Scales in lateral line or body length more than 70. Fig. 565 . . . .

.... **LONGHEAD DARTER,**
Percina macrocephala **(Cope)**

Figure 565.

Pale brown with about 9 confluent spots along each side; small black spot at base of caudal fin. Length 6 inches. Pennsylvania southward on west slope of Appalachians to Georgia.

10b **Scales in lateral line or body length less than 65. Fig. 566** . . . .
. . . . **GILT DARTER,**
Percina evides **(Jordan and Copeland)**

Figure 566.

More or less olivaceous and bronze with about 7 vertically elongated blotches more or less confluent along the lateral line and confluent above with the dorsal blotches forming saddles. Length 4 inches. Eastern Minnesota through the Ohio drainage to New York and south to Oklahoma and northern Georgia.

SHIELD DARTER, *Percina peltata* (Stauffer). Fig. 567. Pale yellowish with dark saddles on back and with about 6 square blotches on the side, these are sometimes confluent. Length 3 inches. Pennsylvania to South Carolina on east side of Appalachians.

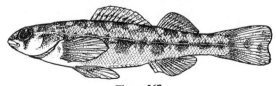

Figure 567.

PIEDMONT DARTER, *Percina crassa* (Jordan and Brayton). Fig. 568. Yellowish with about 6 diffuse and more or less confluent blotches alternating with dark bands on each side. Males with 2 yellow spots at base of caudal fin. Length about 3 inches. Virginia to South Carolina.

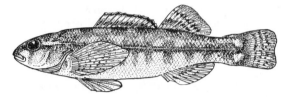

Figure 568.

11a **Scales in lateral line or body length usually more than 65. Fig. 569** . . . .
. . . . **BLACKSIDE DARTER,**
Percina maculata **(Girard)**

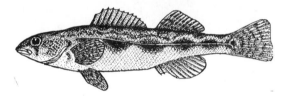

Figure 569.

Pale yellowish with row of 5-6 black and rather elongated blotches on each side and a small black spot at base of caudal fin. Cheeks may have a few scales; belly with strong ridge of scales before anus. Length 3 inches. Southern Manitoba and Ontario to New York and southward west of the Appalachians to northern Alabama and Oklahoma.

LEOPARD DARTER, *Percina pantherina* (Moore and Reeves). Very similar to the blackside darter but has more than 80 scale rows, and is more spotted. Little River System, western Arkansas and eastern Oklahoma.

11b **Scales in lateral line or in body length usually less than 70. Fig. 570** . . . .
. . . . **BLACKBANDED DARTER,**
Percina nigrofasciata **(Agassiz)**

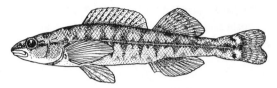

Figure 570.

Olivaceous above and pale below with a row of 10-11 elongated diamond-shape blotches along each side, no bar below the eye. Length 5 inches. Coastal streams, South Carolina to Louisiana.

BRONZE DARTER, *Percina palmaris* (Bailey). Fig. 571. Yellowish brown above and dull olivaceous below with 8-10 dark vertical bars along each side; no dark bar under eye. Coosa-Alabama River System, Alabama and Georgia.

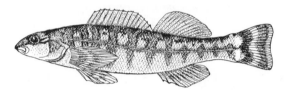

Figure 571.

STRIPEBACK DARTER, *Percina notogramma* (Raney and Hubbs). Markings are similar to those of the blackside darter and the shield darter but has fewer scales in the lateral line and differs in the markings of the back, the mid-dorsal blotches seldom contact laterally with loops as in the related forms. Tributaries of Chesapeake Bay and south to the James River in Virginia.

12a   Cheeks usually naked . . . .
. . . . **PIEDMONT DARTER,**
Percina crassa **(Jordan and Brayton)**
Specimens with gill membranes more closely connected will key here. (See couplet 10b.)

12b   Cheeks more or less scaled **13**

13a   Scales along lateral line or body length less than 70; snout blunt. Fig. 572 . . . . . .
. . . . **DUSKY DARTER,**
Percina sciera **(Swain)**

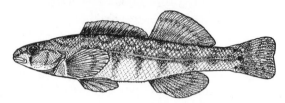

Figure 572.

Yellowish olive; each side with about 7 confluent blotches and with a small black spot at base of caudal fin. Length 5 inches. Indiana southward to Gulf and Texas.

GOLDLINE DARTER, *Percina aureolineata* Suttkus and Ramsey. Similar to dusky darter but more slender and with dorsal saddles indistinct or absent. Cahaba and Coosawattae River drainages of Alabama River system.

RIVER DARTER, *Percina shumardi* (Girard). (See Fig. 558, couplet 4a.) Specimens with an apparent frenum will key here.

13b   Scales along lateral line or in body length (60-80) usually more than 70; snout long and sharp. Fig. 573 . . . .
. . . . **SLENDERHEAD DARTER,**
Percina phoxocephala **(Nelson)**

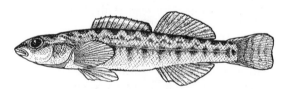

Figure 573.

Yellowish brown with 9-10 quadrate blotches along lateral line, ending in a small black spot at base of caudal fin. Some have gill membranes not so closely connected. Length 6 inches. Minnesota to western Pennsylvania and south to Tennessee and Oklahoma.

FRECKLED DARTER, *Percina lenticula* Richards and Knapp. Snout more blunt than slenderhead darter. Related to dusky darter but has more lateral line scales (77-90) and 3 spots at base of caudal fin tend to fuse with lateral blotches. Gill membranes variously joined. Pearl River eastward into the Alabama River system.

LONGNOSE DARTER, *Percina nasuta* (Bailey). Very similar to the slenderhead darter. Yellowish with 10-14 vertically elongate blotches along lateral line and a small black spot at base of caudal fin; nor dark bar under eye. White River System, Arkansas and Poteau Rivers, Oklahoma.

SHARPNOSE DARTER, *Percina oxyrhyncha* (Hubbs and Raney). Similar in appearance to the slenderhead and the bigheaded darter. Cheat and New Rivers, Virginia and West Virginia.

OLIVE DARTER, *Percina squamata* (Gilbert and Swain). Specimens with belly partly naked will key here. (See couplet 6b.)

GENUS *ETHEOSTOMA*

1a Body depth goes about 7 times in body length; back naked from middle of first dorsal fin forward. Fig. 574. . . . . **GLASSY DARTER,** *Etheostoma vitreum* (Cope)

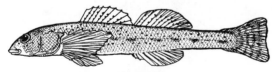

Figure 574.

Body very slender and pellucid; mid-line of belly mostly naked except for a few scales between pelvic fins. Some have only one anal spine. Length 2 inches. Maryland to North Carolina.

1b Body depth goes less than 7 times in body length; back either entirely scaled or naked only in front of first dorsal fin. . . 2

2a Premaxillaries protractile, separated from snout by a complete groove. (Some individuals may have a tiny hidden frenum). 3

2b Premaxillaries not protractile, not entirely separated from snout by a groove, but connected anteriorly by a fleshy bridge 8

3a Anal spines usually 2 4

3b Anal spines usually 1, first soft ray may resemble a spine as it is slender and unbranched, but it is jointed 7

4a Gill membrane only slightly connected across isthmus (Figs. 575, 576). . . . . SPECKLED DARTER, *Etheostoma stigmaeum* (Jordan)

Figure 575.

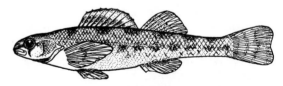

Figure 576.

Olivaceous and speckled with about 8 "W"-shaped blotches on each side. Length 2 1/2 inches. Southeastern Oklahoma, Arkansas, and Tennessee to Georgia and Louisiana.

CHOCTAWHATCHEE DARTER, *Etheostoma davisoni* Hay. Very similar to the speckled darter. Escambia-Choctawhatchee drainages, Alabama.

WACCAMAW DARTER, *Etheostoma perlongum* (Hubbs and Raney). Closely resembles the johnny darter, but is more slender and has over 60 lateral line scales. Length 3 inches. Lake Waccamaw, Columbus County, North Carolina.

4b Gill membranes closely united across isthmus (Fig. 577) 5

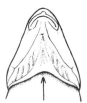

Figure 577.

5a Snout broadly rounded and bulges anteriorly, overhanging the premaxillaries or upper jaw. Fig. 578 . . . .
. . . . GREENSIDE DARTER,
Etheostoma blennioides **Rafinesque**

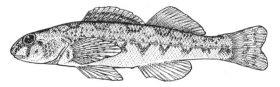

Figure 578.

Body olivaceous; sides marked with series of ''U''-shaped blotches. Length 4 inches. Michigan and Illinois to Pennsylvania and south to Alabama and Oklahoma. Tuskaseegee River, North Carolina.

TENNESSEE SNUBNOSE DARTER, *Etheostoma simoterum* (Cope). Fig. 579. Light green above, yellowish below, with more or less confluent blotches along lateral line. Length 3 inches. Headwaters of Cumberland and Tennessee Rivers.

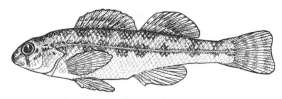

Figure 579.

BLACKSIDE SNUBNOSE DARTER, *Etheostoma duryi* Henshall. Fig. 580. Black greenish above, belly pale or orange; sides with 9-10 quadrate blotches along lateral line. Very similar to Tennessee snubnose darter except in minor pigmentation, suboccular bar tends to curve forward instead of backward. Length 2-3 inches. Upper Tennessee River drainage.

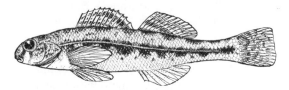

Figure 580.

CUMBERLAND SNUBNOSE DARTER, *Etheostoma atripinne* (Jordan). Fig. 581. Body olivaceous with about 11 bar-like blotches along each side. Length 3 inches. Headwaters of the Green, Cumberland, and Tennessee Rivers.

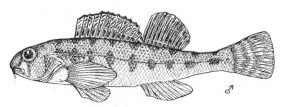

Figure 581.

5b Snout more or less pointed and not greatly overhanging 6

6a Snout shorter than diameter of eye. Fig. 582 . . . .
. . . . RIVERWEED DARTER,

Etheostoma podostemone **Jordan and Jenkins**

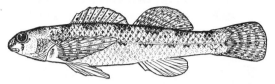

Figure 582.

Grayish brown with sides marked by row of "W"-shaped blotches. Length 3 inches. Headwaters of Roanoke River, Virginia.

6b Snout as long as the diameter of eye. Fig. 583 . . . .
. . . . **LONGFIN DARTER,**
Etheostoma longimanum **Jordan**

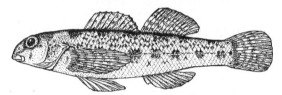

Figure 583.

Grayish brown with sides marked by longitudinal row of "W"-shaped blotches. Length 3 inches. Headwaters of James River, Virginia.

7a Lateral line complete; bar extending forward from eye is broken on tip of snout. Fig. 584 . . . .
. . . . **JOHNNY DARTER,**
Etheostoma nigrum **Rafinesque**

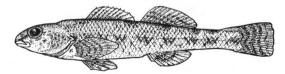

Figure 584.

Grayish brown with sides marked by longitudinal row of "W"-shaped blotches. Length 3 inches. Several sub-species, widespread through upper Mississippi valley, Great Lakes drainage, and Atlantic drainage south to northern Florida.

TESSELLATED DARTER, *Etheostoma olmstedi* Storer. Fig. 585. Related and similar to the johnny darter. Differs in having 13 or more dorsal rays instead of 13 or less. Lake Ontario drainage southward to North Carolina.

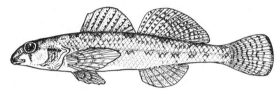

Figure 585.

7b Lateral line is incomplete, extending only a short distance; bar extending forward from eye is continuous around snout. Fig. 586 . . . .
. . . . **BLUNTNOSE DARTER,**
Etheostoma chlorosomum **(Hay)**

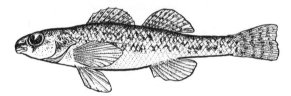

Figure 586.

Grayish brown, sides marked with longitudinal row of "W"-shaped blotches. Length 3 inches, southern Minnesota and Indiana to Alabama and Texas.

8a Usually only one anal spine. Fig. 587. . . .
. . . . **TUSCUMBIA DARTER,**
Etheostoma tuscumbia **Gilbert and Swain**

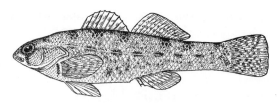

Figure 587.

Greenish with sides mottled or heavily speckled and with 8-10 blotches along lateral line. Some may have 2 anal spines. (See couplet 31b.) Length about 2 inches. Tributaries of upper Tennessee River in Alabama and Tennessee.

TRISPOT DARTER, *Etheostoma trisella* Bailey and Richards. Related to Tuscumbia darter has 3 prominent dorsal blotches and the lateral line is complete (47 scales). Status is uncertain as known only from 2 specimens from Coosa River drainage, Cherokee County, Alabama.

8b Usually 2 anal spines 9

9a Lateral line present, may be complete or incomplete. 10

9b Lateral line entirely or almost absent, 0-7 pores 35

10a Lateral line complete or lacking only on the last 5 or 6 scales 11

10b Lateral line incomplete, usually not extending beyond posterior end of soft dorsal fin. 24

11a Gill membranes more or less closely united across isthmus. (See Fig. 577.) . . .
. . . . 12

11b Gill membranes not united or very slightly united across isthmus. (See Fig. 575.). . 19

12a Interspace between pelvic fins less than half the width of a fin base. Fig. 588 . . . .
. . . . **GOLDSTRIPE DARTER,** *Etheostoma parvipinne* **Gilbert and Swain**

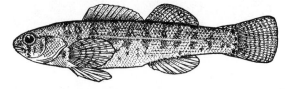

Figure 588.

Olivaceous, sides with about 10 crossbars, usually speckled and with 3 or 4 distinct spots across base of caudal fin. Some individuals with only one anal spine. Length 2 1/2 inches. Tennessee, Alabama and western Florida west to Oklahoma and eastern Texas.

12b Interspace between pelvic fins half or more than the width of a fin base 13

13a Dorsal spines 12 or more. Fig. 589 . . . . . .
. . . . **VARIEGATE DARTER,** *Etheostoma variatum* **Kirtland**

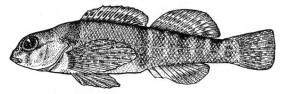

Figure 589.

Dusky green above, light below; sides with 6 or 7 crossbars posteriorly. Spring males very brilliant with vertical orange bars on sides and orange on belly. Length 3 1/2 inches. Ohio River drainage, exclusive of upper Tennessee, Kanawha, Wabash, and Kentucky River systems.

FINESCALE SADDLED DARTER, *Etheostoma osburni* (Hubbs and Trautman). Fig. 590. Dusky green above, light below with 9-11 blotches or bars on sides. Similar to variegate darter. Length 4 inches. Upper Kanawha River drainages, Virginia and West Virginia.

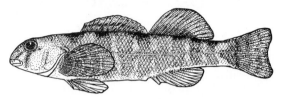

Figure 590.

KANAWHA DARTER, *Etheostoma kanawhae* (Raney). Similar to variegated darter, but breast is naked. Has fewer scales in lateral line (48-57) than finescale saddled darter (59-70). Length 3 1/2 inches. New River drainage, Virginia and West Virginia.

ARKANSAS SADDLED DARTER, *Etheostoma euzonum* (Hubbs and Black). Fig. 591. Light olivaceous brown with 4 dark saddles extending down to lateral line. Very similar to Missouri saddled darter. (See couplet 17b.) Length 3 inches. White River system, southeastern Missouri and northern Arkansas.

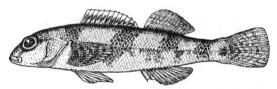

Figure 591.

BLENNY DARTER, *Etheostoma blennius* Gilbert and Swain. Fig. 592. Light olivaceous with dark oblique bars on sides. Length 3 inches. Upper tributaries of Tennessee River.

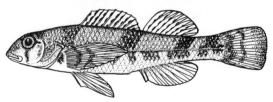

Figure 592.

SWANNANOA DARTER, *Etheostoma swannanoa* Jordan and Evermann. (See Fig. 597, couplet 17b.) Individuals with more than 12 dorsal spines will key here.

13b Dorsal spines usually less than 12 . . . . . **14**

14a Cheeks and opercles, both or only one scaled **15**

14b Cheeks and opercles practically naked, may have 3-5 scales **16**

15a Cheeks naked, opercle scaled. Fig. 593 . . . . **SAVANNAH DARTER,** *Etheostoma fricksium* **Hildebrand**

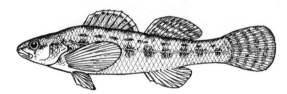

Figure 593.

Brownish above, sides with dark blotches. Length 2 1/2 inches. Savannah River, South Carolina and Georgia.

CHRISTMAS DARTER, *Etheostoma hopkinsi* (Fowler). Closely related to Savannah darter. Olive brown to pale below with 9-10 vertical blotches or streaks on sides. Variable scalation of nape, cheek and opercle. Snout shorter than diameter of eye. Altamaha, Ogeechee, and Savannah River systems.

YOKE DARTER, *Etheostoma juliae* Meek. (See Fig. 634, couplet 34b.) Individuals with rather complete lateral line will key here.

15b Cheeks and opercles usually entirely scaled. Fig. 594 . . . . **BANDED DARTER,** *Etheostoma zonale* (Cope)

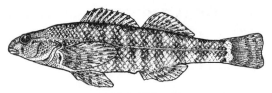

Figure 594.

Olivaceous above, yellowish below; sides with brownish spots along lateral line and 8 narrow bands encircling belly. Length 3 inches. Southern Minnesota to Ohio and western New York and south to Alabama and Arkansas and eastern Oklahoma.

TENNESSEE SNUBNOSE DARTER, *Etheostoma simoterum* (Cope). (See Fig. 579, couplet 5a.) Individuals with rather complete lateral line will key here.

16a Scale rows in lateral line 48 or more . . . 17

16b Scale rows in lateral line less than 48 . . 18

17a Anal soft rays 7-8. Fig. 595 . . . .
. . . . **HARLEQUIN DARTER,** *Etheostoma histrio* **Jordan and Gilbert**

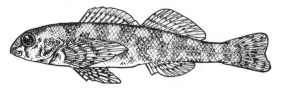

Figure 595.

Dark green with alternating blotches along sides; belly naked. Length 2 inches. Southern Indiana and Kentucky to eastern Texas and western Florida.

ROCK DARTER, *Etheostoma rupestre* Gilbert and Swain. Fig. 596. Greenish; sides marked with numerous small blotches, sometimes "W"-shaped, above and below lateral line. Length 2 inches. Black Warrior River System, Alabama.

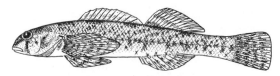

Figure 596.

MARYLAND DARTER, *Etheostoma sellare* (Radcliffe and Welch). Dark lateral blotches (4-5) alternating with 4-5 dark saddles; cheeks and opercle scaly but head, nape, breast and belly naked. Status uncertain as only reported twice. Swan Creek near Aberdeen, Maryland.

17b Anal soft rays 9 or more. Fig. 597 . . . . . .
. . . . **SWANNANOA DARTER,** *Etheostoma swannanoa* **Jordan and Evermann**

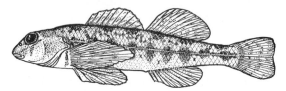

Figure 597.

Dusky green, sides more or less mottled; 8-9 blotches along lateral line. Length 3 inches. Upper tributaries of the Tennessee River.
MISSOURI SADDLED DARTER, *Etheostoma tetrazonum* (Hubbs and Black). Very similar to Arkansas saddled darter (Fig. 591, couplet 13a) but dorsal spines are 12 or less and fewer lateral line scales (49-51). Light olivaceous brown with 4 dark saddles extending down to lateral line and with 9-10 blotches along the side. Big Niangua River and Gasconade River, Missouri.

18a Pectoral fins no longer than head. Fig. 598 . . . .
. . . **SEAGREEN DARTER,** *Etheostoma thalassinum* **(Jordan and Brayton)**

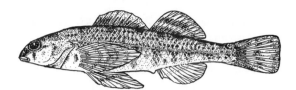

Figure 598.

Dull greenish above and light below, sides with blotches. Males with 6-9 blue green vertical bars on side. Length 2 1/2 inches. Santee River drainage, North and South Carolina.

TURQUOISE DARTER, *Etheostoma mariae* (Fowler). Fig. 599. Dull brownish above and light below; all fins reddish. Length 2 3/4 inches. Cape Fear River drainage, North Carolina.

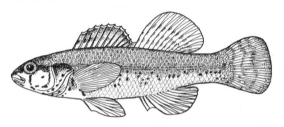

Figure 599.

PINEWOOD DARTER, *Etheostoma inscriptum* (Jordan and Brayton). Fig. 600. Olivaceous with about six blotches along each side. Males with a red spot on each scale, giving appearance of longitudinal lines. Length 2 1/2 inches. Georgia.

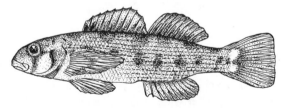

Figure 600.

18b Pectoral fins much longer than head ....
...... BLENNY DARTER, *Etheostoma*
***blennius* Gilbert and Swain**
Light olivaceous with dark oblique bars on sides. Length 3 inches. Upper tributaries of Tennessee River. (See Fig. 592, couplet 13a.)

19a Humeral region with large black humeral
scale. Fig. 601 ....
.... GREENBREAST DARTER,
Etheostoma jordani **Gilbert**

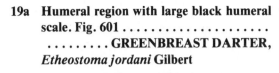

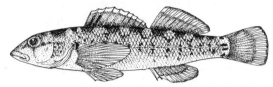

Figure 601.

Olivaceous with 8-10 crossbars on each side. Male with numerous longitudinal lines on sides similar to redline darter. Length 3 inches. Alabama River system in Georgia and Alabama.

REDFIN DARTER, *Etheostoma whipplei* (Girard). (See Fig. 626, couplet 33a.) Individuals with almost complete lateral line will key here.

19b Humeral region without an enlarged
black scale, may have an enlarged but not
very dark scale or a faint spot. 20

20a Snout short and abruptly decurved.
Fig. 602 . . . .
. . . . BLUEBREAST DARTER,
Etheostoma camurum **(Cope)**

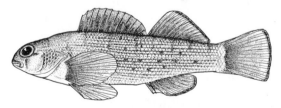

Figure 602.

Brownish and green, vertical fins edged with black. Males with rich blue breasts and with red dots on sides. Length 2 1/2 inches. Ohio River drainage to North Carolina.

ORANGEFIN DARTER, *Etheostoma bellum* Zorach. Similar to the redline darter, but with a prominent sub-orbital bar present, the bar not subdivided. The nape with 1-5 rows of scales posteriorly. Green River system, Kentucky and Tennessee.

BAYOU DARTER, *Etheostoma rubrum* Raney and Suttkus. Small darter similar to bluebreast darter but with saddles and 9-10 vertical bars as in redline darter. Known only from Bayou Pierre and White Oak Creek, Copiah county, Mississippi.

COOSA DARTER, *Etheostoma coosae* (Fowler). Fig. 603. Light brown above and light below; sides mottled and with irregular vertical bands or blotches along lateral line. Length 2 inches. Coosa River, Georgia and Alabama.

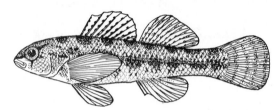

Figure 603.

COLDWATER DARTER, *Etheostoma ditrema* Ramsey and Suttkus. Snout somewhat decurved and may key here. (See couplet 22a.)

20b Snout longer or as long as diameter of eye, not decurved. 21

21a Scale rows in body length more than 70. Fig. 604. . . . .
. . . . NIANGUA DARTER,
Etheostoma nianguae **Gilbert and Meek**

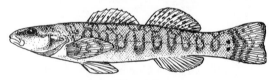

Figure 604.

Olivaceous with 8-9 "U"-shaped bars on each side, 2 black spots at base of caudal fin. Length

4 inches. Niangua and Gasconade River drainages, Missouri.

21b Scale rows in body length less than 70 22

22a Scale rows in body length less than 55. Fig. 605. . . . .
. . . . REDLINE DARTER,
Etheostoma rufilineatum **(Cope)**

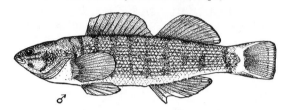

Figure 605.

Greenish with 8 faint crossbars; several horizontal bars on cheek. Males with spots on each scale forming series of longitudinal lines along sides (Fig. 605). Females not marked as strongly (Fig. 606). Length 3 inches. Tributaries of upper Tennessee, Cumberland and Green Rivers.

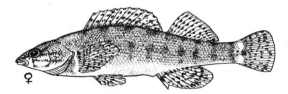

Figure 606.

YELLOWCHEEK DARTER, *Etheostoma moorei* Raney and Suttkus. Similar to bluebreast darter and redline darter but with subocculer bar extending to isthmus cheek with only a single spot behind eye; no red spots on male but some on female; female with dark stripe in center of spiny dorsal fin. Lateral line scales 53-57. Devils Fork and Little Red River in White River system, Arkansas.

RAINBOW DARTER, *Etheostoma*

caeruleum Storer. (See Fig. 620, couplet 30b.) Individuals with rather complete lateral lines will key here.

SAVANNAH DARTER, *Etheostoma fricksium* Hildebrand. (See Fig. 593, couplet 15a.) Individuals with gill membranes very slightly connected will key here.

MUD DARTER, *Etheostoma asprigene* (Forbes). Fig. 607. Brownish above with 9 squarish bar-like blotches on each side. Lateral line usually incomplete. Individuals with a more complete lateral line will key here. Length 2 inches. Minnesota to Indiana and south to Mississippi, northern Louisiana and Texas.

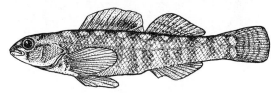

Figure 607.

CREOLE DARTER, *Etheostoma collettei* Birdsong and Knapp. Fig. 608. Similar to the mud darter. Differs in having a dark sub-orbital bar. Humeral spot present. Eight or 9 dorsal saddles, 4 are prominent, one anterior to spinous dorsal, one at posterior end of spinous dorsal, one below soft dorsal and one just behind soft dorsal. Lateral line incomplete, 44-60 in the lateral series, 8-15 without pores. Nape, belly and cheek scaled; breast naked. Length 2 3/4 inches. Little, Red and Sabine drainages in Louisiana and Ouachita River in Louisiana and Arkansas.

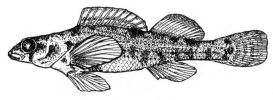

Figure 608.

BLUESIDE DARTER, *Etheostoma jessiae* (Jordan and Brayton). Fig. 609. Similar to mud darter but more slender and with 5-7 dorsal hourglass saddles with narrow part mid-dorsal. Lateral line incomplete with about 35 scales. Upper Tennessee River drainage in Tennessee and Georgia.

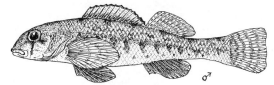

Figure 609.

GULF DARTER, *Etheostoma swaini* (Jordan). Fig. 610. Resembles mud darter and may be a sub-species but lateral line is complete or almost complete and scales are more elevated. Scale spots form numerous irregular longitudinal streaks on sides. Length 2 1/2 inches. Gulf drainage, eastern Louisiana to Florida.

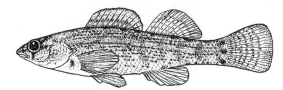

Figure 610.

WATERCRESS DARTER, *Etheostoma nuchale* Howell and Caldwell. Resembles the Gulf darter but differs in fewer dorsal spines and rays and in incomplete lateral line (12-24 pored scales). Reported only from Glen Spring, Bessemer, Alabama.

COLDWATER DARTER, *Etheostoma ditrema* Ramsey and Suttkus. Brownish and mottled with 3 black spots at base of caudal fin. Snout rather decurved and less than diameter of eye; cheeks and opercle scaled. Scales in body length 41-54; only 19-35 scales in lateral line. Dorsal fin rays 8-12. Springs and spring-ponds, Coosa-Alabama River system, Georgia and Alabama.

22b Scale rows in body length more than 55 . . . . **23**

23a Sides marked with blotches and with more or less fine horizontal lines. Fig. 611 . . . .
. . . . SPOTTED DARTER, *Etheostoma maculatum* Kirtland

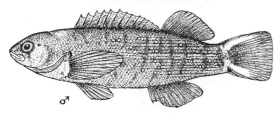

Figure 611.

Dark olivaceous with a blue throat; scales on sides with spots giving appearance of numerous fine lines. Cheeks with several horizontal bars. Fins speckled in female. Length 2 1/2 inches. Ohio to northern Alabama.

SMALLSCALE DARTER, *Etheostoma microlepidum* Raney and Zorach. Closely related to the yellowcheek, redline and spotted darter but has 13 dorsal spines and more scales in the lateral series (60-70). Belly partly naked. Resembles the redline darter as caudal fin of male is marked the same but the spiny dorsal fin has a light margin and the soft dorsal fin has a thin black margin. The snout is more pointed in the smallscale darter. Female not brightly colored but strongly blotched. Cumberland River below falls, Duck River of the Tennessee River drainage, Kentucky and Tennessee.

SHARPHEAD DARTER, *Etheostoma acuticeps* Bailey. Dark longitudinal streaks on sides as in spotted darter; soft anal and dorsal fins without black margins; 10 dorsal spines; snout very sharp and longer than diameter of eye. Sub-opercular bar faint or absent. Lateral line scales 54-61. Opercle naked. South Fork Holston River, upper Tennessee River drainage.

ASHY DARTER, *Etheostoma cinereum* Storer. Fig. 612. Yellowish brown with 10-12 dark oblong blotches on sides. Scale spots form several longitudinal lines above lateral line. Length 4 inches. Tributaries of upper Tennessee River.

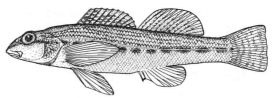

Figure 612.

23b Sides marked with "U"-shaped bars. Fig. 613 . . . .
. . . . . . ARROW DARTER, *Etheostoma sagitta* (Jordan and Swain)

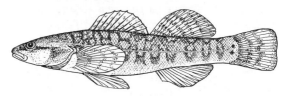

Figure 613.

Dusky green with about 9 "U"-shaped bars on sides. Length 3 inches. Tributaries of Kentucky River and upper Cumberland River, Kentucky.

24a Lateral line rather elevated anteriorly; not over 3 scale rows between lateral line and base of first dorsal fin 25

24b Lateral line more or less straight; 4 or more scale rows between lateral line and base of first dorsal fin 26

25a Snout rather blunt; body elongate and slender. Fig. 614 . . . .
. . . . SLOUGH DARTER, *Etheostoma gracile* (Girard)

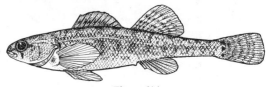

Figure 614.

Olivaceous with sides mottled and marked with about 9 blotches; distance from tip of snout to angle of gill membranes about 1/2 length of head. Base of caudal fin with 3 vertical spots. Length 2 inches. Southern Illinois and Indiana to Mississippi and central Texas.

SWAMP DARTER, *Etheostoma fusiforme* (Girard). Fig. 615. Olivaceous with blotches along each side which may fuse to form an irregular lateral band; 3 or 4 spots across base of caudal fin; distance from tip of the snout to the angle of the gill membranes decidedly greater than 1/2 the length of the head. Widespread from Massachusetts to Maryland and represented by sub-species *E.f. barratti* (Holbrook) (Fig. 616) from Virginia to eastern Texas.

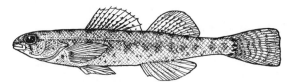

Figure 615.

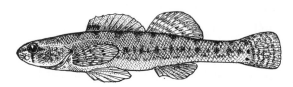

Figure 616.

BACKWATER DARTER, *Etheostoma zoniferum* (Hubbs and Cannon). Similar to slough darter, but gill membranes more closely connected; distance from the tip of snout to angle of gill membranes somewhat greater than 1/2 the length of head; sides marked anteriorly with irregular blotches which form about 4 vertical bars on the peduncle; 3 spots across base of caudal fin. Central part of Alabama River system.

SAWCHEEK DARTER, *Etheostoma serriferum* (Hubbs and Cannon). Fig. 617. Similar to the swamp darter, but the lateral line is longer, reaching to below middle of soft

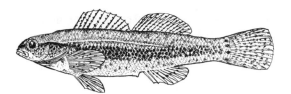

Figure 617.

dorsal fin; gill membranes slightly closely connected; distance from tip of snout to angle of the gill membranes is greater than 1/2 the length of the head; opercle more saw toothed. Sides marked with more or less confluent blotches, 4 spots across base of caudal fin. South and North Carolina and southeastern Virginia.

BROWN DARTER, *Etheostoma edwini* (Hubbs and Cannon). (See Fig. 622, couplet 31b.) Individuals with only 3 rows of scales above lateral line will key here.

25b **Snout rather pointed; body stout. . . . . . .**
. . . . **SALUDA DARTER,**
Etheostoma saludae **(Hubbs and Cannon)**
Similar to swamp darter. About 10 rectangular blotches along each side and with 3-4 spots across base of caudal fin. Santee River system in South Carolina.

CAROLINA DARTER, *Etheostoma colle* (Hubbs and Cannon). Sides marked with median dark stripe breaking into blotches on the peduncle; 3 dark spots across base of caudal fin. Catawba River system, South Carolina.

26a **Gill membranes scarcely connected across isthmus. (See Fig. 575.) 27**

26b **Gill membranes more or less connected across isthmus. (See Fig. 577.) 34**

27a **Shoulder region without a black humeral scale, may have a faint dark spot 28**

27b **Shoulder region wtih a distinctly enlarged black humeral scale 32**

28a Opercle scaleless; cheek may or may not be scaled. Fig. 618 . . . . **GREENTHROAT DARTER,** *Etheostoma lepidum* **(Baird and Girard)**

Figure 618.

Olivaceous, sides marked with bars. Very similar to rainbow darter, but head is more naked. Length 2 1/2 inches, Oklahoma south through central Texas.

ORANGETHROAT DARTER, *Etheostoma spectabile* (Agassiz). Fig. 619. Olivaceous with about 10 dark bars on sides. Length 2 1/2 inches. Eastern Colorado to Ohio and Tennessee.

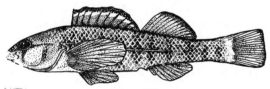

Figure 619.

28b Opercle with scales; cheek may or may not be scaled 29

29a Cheeks nearly naked; opercles scaled . . 30

29b Cheeks and opercles more or less scaled 31

30a Gill membranes not connected across isthmus . . . . **RIO GRANDE DARTER,** *Etheostoma grahami* **(Girard)**

Resembles greenthroat darter except opercle is scaled and lateral line has more than 50 scales. Males have throat red instead of blue. Rio Grande drainage of Texas and Mexico.

ORANGETHROAT DARTER, *Etheostoma spectabile* (Agassiz). Some subspecies have the opercle completely scaled and will key here. (See Fig. 619, couplet 28a.)

30b Gill membranes very slightly connected across isthmus. Fig. 620 . . . . **RAINBOW DARTER,** *Etheostoma caeruleum* **Storer**

Figure 620.

Olivaceous with 10-12 dark bars on each side. Breeding males very brilliant. Length 2 1/2 inches. Southern Minnesota to eastern Ontario and south to Alabama and Arkansas.

TIPPECANOE DARTER, *Etheostoma tippecanoe* Jordan and Evermann. Individuals without a dark humeral scale will key here. (See couplet 32a.)

31a Lateral line extending on 47 to 53 scales . . . . **MUD DARTER,** *Etheostoma asprigene* **(Forbes)**

Individuals with a rather complete lateral line will key here. (See Fig. 607, couplet 22a.)

GULF DARTER, *Etheostoma swaini* (Jordan). (See Fig. 610, couplet 22a.) Individuals with a more complete lateral line will key here.

SWAMP DARTER, *Etheostoma fusiforme* (Girard). (See Fig. 615, couplet 25a.) Individuals with more than 3 rows of scales above lateral line will key here.

BLUESIDE DARTER, *Etheostoma jessiae* (Jordan and Brayton). (See Fig. 609, couplet 22a.) Individuals with more complete lateral line will key here.

31b **Lateral line very short, usually not extending far behind first dorsal fin and extending on 13 to 40 scales. Fig. 621. . . . . .IOWA DARTER,** *Etheostoma exile* **(Girard)**

Figure 621.

Pale olivaceous with 9-11 blotches or vertical bars on each side. Spawning males very brilliant with red spots on sides and blue and red bands acrss dorsal fins. Length 2 1/2 inches. Widely distributed from southern Canada and Great Lakes drainage through the upper Mississippi valley from Colorado to Ohio.

BROWN DARTER, *Etheostoma edwini* (Hubbs and Cannon). Fig. 622. Grayish brown with 8 or 9 spots along lateral line. Length 2 inches. Southeastern Alabama, Florida, and southern Georgia.

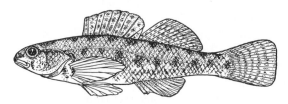

Figure 622.

OKALOOSA DARTER, *Etheostoma okaloosae* (Folwer). Resembles brown darter but has lateral line scales 34-35 instead of 36-40 and differs slightly in markings; first anal spine is longer rather than shorter than second anal spine. Okaloosa and Walton Counties, Florida.

TUSCUMBIA DARTER, *Etheostoma tuscumbia* Gilbert and Swain. Individuals considered as having 2 anal spines will key here. (See Fig. 587, couplet 8a.)

SPOTTAIL DARTER, *Etheostoma squamiceps* Jordan. Fig. 623. Dusky olive with

sides mottled sometimes forming about 10 irregular bars. Length 2 inches. Humeral scale present, but may not be very dark. Indiana southward through western Kentucky and Tennessee to Mississippi.

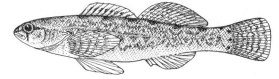

Figure 623.

REDBAND DARTER, *Etheostoma luteovinctum* Gilbert and Swain. Fig. 624. Lateral line on 30-35 scales. Pale olive; 7 dusky bars on back reaching to lateral line and 9 dusky blotches on each side below lateral line; black spots at base of caudal fin. Length 2 inches. Upper Tennessee River drainage.

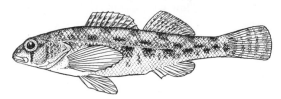

Figure 624.

32a **Cheeks almost naked; opercles scaled . . . . . TIPPECANOE DARTER,** *Etheostoma tippecanoe* **Jordan and Evermann.**
Body dusky, marked by fine dots and by 12 crossbars on each side; belly with naked midline. Length 1 1/2 inches. Ohio River system exclusive of the Tennessee River basin.

SPOTTAIL DARTER, *Etheostoma squamiceps* Jordan. (See Fig. 623, couplet 31b.) Individuals with a dark humeral scale will key here.

ARKANSAS DARTER, *Etheostoma cragini* Gilbert. (See Fig. 625, couplet 33a.) Individuals with a few scales on cheeks and opercles will key here.

32b **Cheeks and opercles without scales . . . 33**

33a Snout shorter than diameter of eye. Fig. 625. . . . . **ARKANSAS DARTER,** *Etheostoma cragini* **Gilbert**

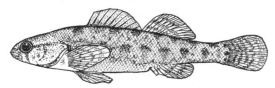

Figure 625.

Olivaceous and mottled; speckled below lateral line; 10-12 dusky spots along side; humeral scale black and conspicuous. Length 2 inches. Arkansas River drainage Colorado to Arkansas.

PALEBACK DARTER, *Etheostoma pallididorsum* Distler and Metcalf. Upper sides dark brown with 1-6 faint vertical bars; dark humeral scale. Male with orange on cheeks and opercle; an orange and a black bar on spiny dorsal fin. Related to Arkansas darter but differs in some markings, particularly by a light mid-dorsal streak with 4-5 indistinct saddles. Caddo River, Ouachita River system, Arkansas.

REDFIN DARTER, *Etheostoma whipplei* (Girard). Fig. 626. Grayish and mottled. Males with red spots on sides and with indistinct bars posteriorly. Black humeral scale present. Length 2 1/2 inches. Lower Arkansas River drainage, Red River in Texas east to Alabama.

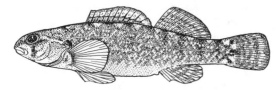

Figure 626.

ORANGEBELLY DARTER, *Etheostoma radiosum* (Hubbs and Black). Very similar to redfin darter, but differs in having more than 62 scale rows instead of less than 61 scales.

Southeastern Oklahoma and southern Arkansas.

SLACKWATER DARTER, *Etheostoma boschungi* Wall and Williams. Olivaceous to brownish, blue black sub-orbital bar present. Three prominent dorsal saddles present. Scales in lateral series 43-58, 6-19 unpored. Tennessee River in northern Alabama and western Tennessee.

33b Snout as long or longer than diameter of eye. Fig. 627. . . . . **STIPPLED DARTER,** *Etheostoma punctulatum* **(Agassiz)**

Figure 627.

Dark green; sides finely dotted and mottled. Males with indistinct bars on sides. Length 2 inches. Ozark region of Missouri, Arkansas, and Oklahoma.

BARCHEEK DARTER, *Etheostoma obeyense* Kirsch. Fig. 628. Related to fantail darter. Light olive with 10-11 lateral blotches forming bars on each side; large humeral scale present. Length 2 1/2 inches. Upper Cumberland River drainage, Kentucky and Tennessee.

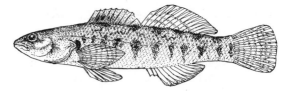

Figure 628.

TEARDROP DARTER, *Etheostoma barbouri* Kuehne and Small. Fig. 629. Similar to the preceding two species but differs in having a prominent sub-orbital bar. Coloration dull to bright yellow with brownish or blackish lateral blotches. Faint horizontal lines usually present on sides. Lateral line incomplete, 40-49 scales in lateral series, usually less than 9 pored scales (0-12). Green River basin of Kentucky and Tennessee.

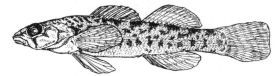

Figure 629.

STRIPED DARTER, *Etheostoma virgatum* (Jordan). Fig. 630. Grayish green with a spot on each scale forming series of longitudinal lines; lateral line does not reach soft dorsal. Length 2 1/2 inches. Cumberland River drainage, Kentucky and Tennessee.

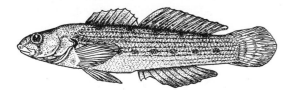

Figure 630.

34a Lateral line reaching only to front of soft dorsal fin; spinous dorsal fin very low, only about 1/2 height of soft dorsal fin. Fig. 631. . . . . FANTAIL DARTER, *Etheostoma flabellare* **Rafinesque**

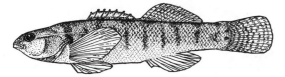

Figure 631.

Grayish green; body scales each with a spot, sometimes forming longitudinal lines (*E. f. lineolatum* Agassiz, Fig. 632); large black humeral scale present; sides with 9-11 crossbars. Lower jaw more or less projecting. Dorsal spines of male with knobs. Length 3 inches.

Several sub-species from Minnesota to Vermont and south to North Carolina and Oklahoma.

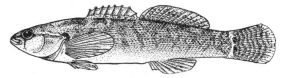

Figure 632.

STRIPETAIL DARTER, *Etheostoma kennicotti* (Putnam). Fig. 633. Resembles fantail darter, but lacks dots, lines, and bars on sides. Upper Cumberland River and adjacent Tennessee River drainages.

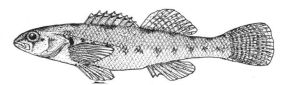

Figure 633.

34b Lateral line almost complete or reaching past front of soft dorsal fin; spinous dorsal fin more than 1/2 the height of the soft dorsal fin. Fig. 634 . . . . YOKE DARTER, *Etheostoma juliae* **Meek**

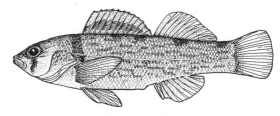

Figure 634.

Dusky olive to greenish; faint yellowish scale dots form irregular longitudinal lines on sides; 5 or 6 faint bars on each side. Length 2 1/2 inches. Southern drainage (White River) of the Ozarks, Missouri.

35a Anal spines 1. Fig. 635. . . . .
. . . . FOUNTAIN DARTER,
Etheostoma fonticola (Jordan and
Gilbert)

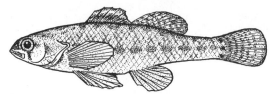

Figure 635.

Olivaceous and somewhat mottled with a
median row of blotches on each side; 3 spots
across base of caudal fin. Length 1 1/2 inches.
Central Texas.

35b Anal spines 2 36

36a Cheeks and opercles mostly naked
(several scales may be on opercle).
Fig. 636. . . . .

. . . . . . LEAST DARTER, *Etheostoma
microperca* Jordan and Gilbert

Figure 636.

Olivaceous mottled with brown; pelvic fins very
long. Length 1 1/2 inches. Southeastern
Manitoba and Great Lakes region to Kentucky
and Oklahoma.

36b Cheeks and opercles completely scaled . .
. . . . CYPRESS DARTER,
Etheostoma proeliare (Hay)
Very similar to the least darter; may have
several pored scales representing lateral line.
Length 1 1/2 inches. Western Tennessee and
Arkansas southward to eastern Texas and
western Florida.

DRUM FAMILY
Sciaenidae

The drum family contains mostly marine
species, many of which are important game
fishes and may occur in the mouths of rivers
along the coasts. The family contains but one
strictly feshwater species, the freshwater drum,
Aplodinotus grunniens Rafinesque (Fig. 637)
which is widespread, extending from the James
Bay drainage in Canada through the Great
Lakes drainage exclusive of Lake Superior and
southward into Mexico. This is a deep bodied
silvery fish with a high back and a long dorsal
fin. The lateral line extends out onto the caudal
fin.

Like many of the marine members of this
family, the freshwater drum makes a rumbling

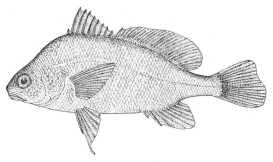

Figure 637.

noise supposed to be produced by the swim
bladder. It feeds mostly on the bottom, grinding its food with a powerful set of flat

pharyngeal teeth. It possesses large ivory-like ear bones or ossicles within its skull which are quite unlike those of any other freshwater fish. The freshwater drum reaches a size of 10 pounds and a length of several feet, although much larger sizes were reported in the past century.

Along the Atlantic coast a large number of marine species may enter brackish and fresh water at the mouths of rivers. The red drum, *Sciaenops ocellata* (Linnaeus); the spot, *Leiostomus xanthurus* Lacepede; the black drum, *Pogonias cromis* (Linnaeus); the croaker, *Micropogon undulatus* (Linnaeus); the spotted seatrout, *Cynoscion nebulosus* (Valenciennes); and several others have been reported from fresh and brackish waters along the Atlantic coast.

SURF-FISH FAMILY
Embiotocidae

The surf fishes form a large marine group with one species living in fresh water in California. The family is differentiated by numerous characters, of which one of the most outstanding is a sheath of scales separated from the body by a groove and extending along the base of the spiny and soft dorsal fins. The members of this family are live bearers, retaining the eggs in the ducts of the female until the young are fully developed.

The tuleperch, *Hysterocarpus traski* Gibbons (Fig. 638) occurs in the streams of central California, mostly in the drainage of the Sacramento River. It is a small deep bodied fish with a bluish back and silvery sides. Three color phases are present: wide bands on the sides, narrow bands, and no bands. The coloration or phrases are not related to sex. The species reaches a length of 4 to 5 inches. It can be

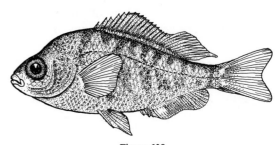

Figure 638.

distinguished from the shiner perch by the presence of 15 or more spines in the dorsal fin.

The shiner perch, *Cymatogaster aggregata* Gibbons, a common marine fish along the Pacific coast is found in fresh water in California coastal streams. It has 10 or less spines in the dorsal fin.

CICHLID FAMILY
Cichlidae

This is a large family of fishes native to rivers and lakes in Africa, Central and South America. One species, the Rio Grande perch, *Cichlasoma cyanoguttata* (Baird and Girard) (Fig. 639) ranges from central America northward into Texas. It has been introduced into other areas. Members of this family are characterized by a broken lateral line and by only one aperature for each nostril. The Rio Grande perch has a deep body which is brownish and is speckled all over, including the fins with blue spots. Black spots may occur on the dorsal fin, back and at the base of the caudal fin. It reaches a length of about 8 inches.

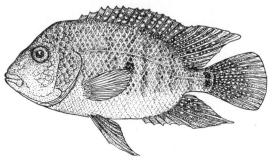

Figure 639.

Cichlids are popular aquarium fishes because of their breeding behavior and their brilliant colors. Many of the species are very hardy and adapt readily to our southern waters. As predators they compete with our native basses and sunfishes. At least eight species have been introduced and become established in North America. These include the South American black acara, *Aequidens portalegensis,* and the oscar, *Astronotus ocellatus,* known from the waters of south Florida. Two species of *Tilapia,* the Mozambique and blackchin mouthbrooder (*T. mossambica* and *T. melanotheron*) have been introduced from Africa. *Tilapia mossambica* occurs in Alabama, Texas and warm springs in Montana; *T. melanotheron* is found in west-central Florida. Two Central American cichlids have also become established; the banded cichlid, *Cichlasoma severum,* in Roger Spring, Nevada and the convict cichlid, *C. nigrofasciatum,* is known from two localities in south Florida, Roger Spring and Ash Springs, Nevada, and a hot spring at Banff, Alberta. At least five other cichlids have been released in Florida and one in Alabama but it is unknown whether any of these have become established.

At least four species have been introduced and established in Hawaii where they reach a large size and are considered as pan fish. These include *Tilapia machochir, T. zilli, T. melanopleura* and *T. mossambica.* The oscar, *Astronotus ocellatus* and tucunare, *Cichla occellaris* from Central and South America have become established and are considered game fishes.

MULLET FAMILY
Mugilidae

This family contains the mullets which are found in the warmer seas over the world. Mullets are characterized by small weak mouths and by a small 4 spined dorsal fin considerably in advance of the soft portion of the dorsal fin. The pectorals are located high on the sides of the body. They may bear conspicuous adipose membranes over their eyes as in the herrings.

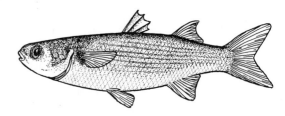

Figure 640.

Several species of mullets enter fresh water in the southern United States. The striped mullet, *Mugil cephalus* Linnaeus, Fig. 640, penetrates up the rivers of the Gulf states and in California. The writer has collected it in the Mississippi river over a hundred miles above the mouth. It may reach a length of 2 feet.

The white mullet, *Mugil curema* Valenciennes also penetrates into river mouths from Cape Cod to Texas. Several other species of mullets may stray into fresh water. The mountain mullet, *Agonostomus monticola* (Bancroft) found in the West Indies, occurs in small streams of eastern Florida and westward to Louisiana. It has a more pointed snout and lacks the adipose eyelids. It also has a darker back and dorsal fins with a spot at the base of the caudal fin.

GOBY FAMILY
Gobiidae

This family includes the gobies and the sleepers (Eleotridae). They are essentially marine fishes found mostly in shallow water along the seashores and some may enter rivers. The pectoral fins are large and the lateral line is absent. The gobies are characterized by a ventral sucker disc formed by the fusion of the pelvic fins (Fig. 641). The pelvic fins of the sleepers are close together but do not unite to form a sucking disc.

The gobies are usually small, 2-4 inches in length but several species may reach a length of over 12 inches. The freshwater goby, *Gobionella shufeldti* (Jordan and Eigenmann) (Fig. 642) occurs in lower stretches of streams emptying into the Gulf of Mexico and the

Figure 641.

Atlantic as far north as Georgia. It may reach a length of 4 inches.

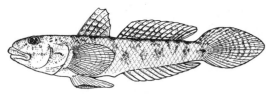

Figure 642.

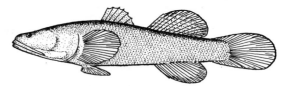

Figure 644.

The river goby, *Awaous tajasica* (Lichtenstein) reaching a length of 12 inches enters streams in Florida on both the Atlantic and Gulf coasts. The marked goby, *Gobionellus stigmaticus* (Poey) (Fig. 643), reaching a length of 5 inches occasionally enters fresh water along the coasts of the southern Atlantic and the Gulf of Mexico.

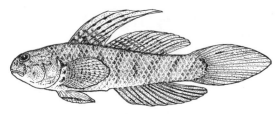

Figure 643.

A number of gobies have been collected from the St. Johns River, Florida, including the violet goby, *Gobiodes braussonneti* Lacepede, the darter goby, *Gobionellus boleosoma* (Jordan and Gilbert), the naked goby, *Gobiosoma bosci* (Lacepede), the code goby, *G. robustum* Ginsburg, the clown goby, *Microgobius gulosus* (Girard), and the green goby, *M. thalassinus* (Jordan and Gilbert). The yellow fin goby, *Acanthogobius flavimanus* (Temmink and Schlegel), has been collected from the delta of the San Joaquin River delta, California.

In southern California, the longjaw mudsucker, *Gillichthys mirabilis* Cooper (Fig. 644), the arrow goby, *Clevelandia ios* (Jordan and Gilbert) and the tidewater goby, *Eucyclogobius newberryi* (Girard) may enter the mouths of rivers. Several freshwater gobies such as *Chonophorus stamineus* (Eydoux and Souleyet)

are native to the streams of Hawaii but spend part of their life history in salt water.

The sleepers include a number of species which occur may enter fresh water close to the sea in the southern states. Several species are common in the rivers of Mexico and the West Indies. The fat sleeper, *Dormitator maculatus* (Bloch) (Fig. 645) occurs in the lower stretches of rivers from South Carolina to Texas into Mexico. It has a short heavy body which is brown with light spots. The maxillary reaches to the anterior margin of the eye. It reaches a length of over 12 inches.

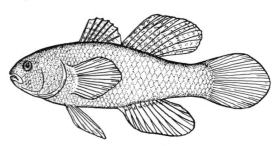

Figure 645.

The spinycheek sleeper, *Eleotris pisonis* (Gmelin) is usually found in brackish and fresh water along the Gulf of Mexico. It has a rather elongated body which is brown and the maxillary reaches to the back of the orbit. It reaches a length of 6-7 inches. A similar form, the spotted sleeper, *Eleotris picta* Kner and Steindachner enters the Colorado River in southern California.

A number of other sleepers and gobies have been reported entering fresh water in southern Florida.

LEFTEYE FLOUNDER FAMILY
Bothidae

The flatfish family contains a number of species on the Atlantic and Pacific coasts, several of which sometimes enter the mouths of rivers along the Atlantic coasts. The bay whiff, *Citharichthys spilopterus* Gunther (Fig. 646), is occasionally found in fresh water from New York to Texas. It is olive brown with dark blotches and reaches a length of 5-6 inches. The jaws are equal on both sides. Another small flatfish occasionally entering fresh water is the fringed flounder, *Etropus crossotus* Jordan and Gilbert distributed from Virginia southward. It is olive brown with dark blotches, but has unequal jaws. The opercle on the blind side has a distinct white fringe along the margin. It reaches a length of almost 6 inches. On the east

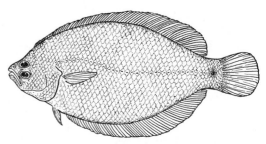

Figure 646.

coast from New York to Texas, the southern flounder, *Paralichthys lethostigma* Jordan and Gilbert is sometimes found in fresh water. It is dusky olive with some darker spots and may reach a large size, 2 to 3 feet in length.

RIGHTEYE FLOUNDER FAMILY
Pleuronectidae

This is a large and abundant marine group widely distributed on both coasts. Several species, especially the young, may enter rivers for a short distance. The starry flounder, *Platichthys stellatus* (Pallas) (Fig. 647) is widely distributed along the Pacific coast and may enter rivers. It is characterized by 4 to 5 prominent vertical black bands in both the dorsal and anal fins. It is commonly 12 to 15 inches in length but may reach a much larger size.

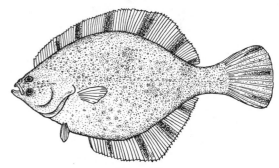

Figure 647.

SOLE FAMILY
Soleidae

This family contains small flatfishes which have the pectoral fins poorly developed, or have one or both fins absent. Several species may enter the mouths of rivers along the Atlantic and Gulf coasts. They are usually only 5 to 6 inches in length. The hogchoker, *Trinectes maculatus* (Bloch and Schneider) (Fig. 648) is common along the Atlantic coast. It is brown with about 8 crossbars and lacks both pectoral fins. Along the southern Atlantic and Gulf coasts, the lined sole, *Achirus lineatus* (Linnaeus) sometimes enters fresh water. It is dark with about 8 crossbars but has a small right pectoral fin.

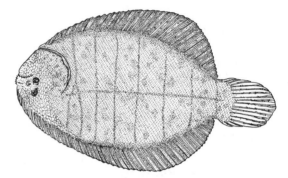

Figure 648.

GUNNEL FAMILY
Pholidae

The Pholidae are blennoid fishes found in the littoral or inter-tidal zone on the Pacific and Atlantic coasts. They have elongate bodies and gill membranes that are joined. The snout is short, the teeth are small and the body is covered with very small scales. Members of the family are marine but two species, the penpoint gunnel, *Apodichthys flavidus* Girard, and the saddleback gunnel, *Pholis ornata* (Girard), have been collected in the Navarro River, Mendocino County, California. The collections were made a short distance above the mouth.

The penpoint gunnel has pelvic fins, its

Figure 649.

coloration is variable ranging from orange to greenish; the anal spine is large, shaped like a pen, and is enclosed in a pouch of skin. The saddleback gunnel lacks pelvic fins and has about ten V-shaped marks or saddles on the back. (Fig. 649)

Some Useful Literature for the Identification of Fishes

Bailey, Reeve M., et alles. 1960. A List of Common and Scientific Names of Fishes from the United States and Canada. Spec. Publ. No. 2, Am. Fish. Soc., 2nd ed., 102 pp. A checklist of the proper common and scientific names of fishes.

Bailey, Reeve M. and M.O. Allum. 1962. Fishes of South Dakota. Misc. Publ. Mus. Zool. Univ. Mich., No. 119, pp. 131. Habits and descriptions of the fishes of South Dakota.

Bailey, R.M. and C.E. Bond. 1963. Four New Species of Freshwater Sculpins, Genus *Cottus,* from Western North America. Occ. Pap. Mus. Zool., Univ. Mich., No. 634, 27 pp.

Berg, Leo S. 1940. Classification of Fishes, Both Recent and Fossil. Trav. Inst. Zool. Acad. Sci. U. R. S. S. Tome 5, Livr. 2: pp. 87-517. A very modern classification of the fishes of the world to families with some illustrations, but no keys.

Bond, Carl E. 1961. Keys to Oregon Freshwater Fishes. Agric. Exp. Sta., Oregon State Univ., Technical Bull. 38, 42 pp.

Brock, V.E. 1960. The Introduction of Aquatic Animals into Hawaiian Waters. Int. Revue ges. Hydrobiol. 45 (4); 463-480. A description of all the fishes introduced into Hawaii up to 1960.

Brown, C.J.D. 1971. Fishes of Montana. Big Sky Books. Montana State University. Bozeman. 207 pp. Habits, descriptions and distribution of the fishes of Montana.

Buchanan, T.M. 1973. Key to the Fishes of Arkansas. Arkansas Game and Fish Comm., 68 pp., 198 maps. Key to the fishes of Arkansas with distribution maps.

Carl, G.C., W.A. Clemens, and C.C. Lindsey. 1959. The Fresh-water Fishes of British Columbia. British Columbia Prov. Mus. Handbook, No. 5, 3rd ed., 192 pp. Keys and illustrations of the fishes of British Columbia. Includes many northwestern fishes which range into Alaska.

Carr, Archie, and C.I. Goin. 1955. Guide to the Reptiles, Amphibians and Fresh-Water Fishes of Florida. Univ. Florida Press, Gainesville. i-viii; 341 pp. Keys, ranges, and descriptions of fishes found in fresh waters of Florida with some illustrations.

Clay, W.M. 1975. The Fishes of Kentucky. Kentucky Dept. of Fish and Wildlife Resources. 416 pp. Habits, descriptions and distribution of the fishes of Kentucky.

Clemens, C.W. and G.V. Wilby. 1961. Fishes of the Pacific Coast of Canada. Bull. Fish. Res. Bd. Canada, No. 68 2nd ed.; 443 pp. Includes many anadromous fishes ranging into Alaska.

Cross, Frank B. 1967. Handbook of Fishes of Kansas. Misc. Publ. No. 45, Univ. Kansas Mus. Nat. Hist.; 357 pp. Habits and descriptions of the fishes of Kansas, illustrated.

Douglas, N.H. 1974. Freshwater Fishes of Louisiana. Claitor's Publ. Div., Baton Rouge, Louisiana. 443 pp. Habits, descriptions and distribution of freshwater fishes of Louisiana.

Eddy, Samuel, and J.C. Underhill. 1974. Northern Fishes with Special Reference to the Upper Mississippi Valley. Univ. of Minn. Press, Mpls., 2nd Ed., 404 pp. Keys and descriptions of the fishes of Minnesota with illustrations.

Fedoruk, A.N. 1969. Checklist of and Key to the Freshwater Fishes of Manitoba. Manitoba

Dept. Mines and Natural Resources, Rept. 6, 99 pp. Key to the fishes of Manitoba with illustrations.

Forbes, S.A. and R.E. Richardson. 1920. The Fishes of Illinois. 2nd Ed. Nat. Hist. Surv. Ill., Vol. 3: cxxi, 357 pp. Atlas, 103 maps. Keys and descriptions of the fishes of Illinois with illustrations. Nomenclature not up to date.

Fowler, H.W. 1945. A study of the Fishes of the Southern Piedmont and Coastal Plain. Acad. Nat. Sci. Philadelphia, Monogr. No. 7, 408 pp.

Gerking, S.D. 1945. The distribution of the Fishes of Indiana. Invest. Indiana Lakes and Streams, 3(1); 1-137.

Gosline, W.A. 1971. Functional Morphology and Classification of Teleostean Fishes. University Press of Hawaii, Honolulu, 208 pp.

Greenwood, P.H., D.E. Rosen, S.H. Weitzman, and G.S. Myers. 1966. Phyletic Studies of Teleotean Fishes, with a Provisional Classification of Living Forms. Bull. Am. Mus. Nat. Hist., 131 (4); 339-456. The most recent study and arrangement of the families of fishes.

Gunter, Gordon. 1942. A list of the Fishes of the Mainland of North and Middle America Recorded from Both Freshwater and Sea Water. Am. Midl. Nat., Vol. 28: pp. 305-356. A list of the marine fishes known to enter fresh water and the freshwater fishes entering salt water. No keys or illustrations.

Harlan, J.R. and E.B. Speaker. 1951. Iowa Fishes and Fishing. Iowa State Cons. Comm., Des Moines. 237 pp. Keys and descriptions of the fishes of Iowa with excellent illustrations.

Hubbs, Carl L., and Karl F. Lagler. 1964. Fishes of the Great Lakes Region. Univ. of Mich. Press, Ann Arbor, 213 pp. Keys and ranges of the fishes of the Great Lakes drainage with illustrations.

Jordan, David S., and Barton W. Evermann. 1896-1900. The Fishes of North and Middle America. Bull. U.S. Nat. Mus., No. 47, in 4 parts, 313 pp. Out of date but still the classical work covering all marine fishes and freshwater fishes of North America known at the time of publication. With keys, ranges, descriptions and some illustrations.

Jordan, D.S., B.W. Evermann, and H.W. Clark. 1930. Checklist of the Fishes and Fishlike Vertebrates of North and Middle America North of the Northern Boundary of Venezuela and Columbia. Rept. U.S. Comm. Fish., 1928, Pt. 2: iv, 670 pp. A complete list of names for all species with synonyms for both marine and freshwater fishes. Many species are invalid and nomenclature is not up to date, but still very useful. No keys or illustrations.

Knapp, Frank T. 1953. Fishes Found in the Fresh Waters of Texas. Ragland Studio and Litho. Printing Co., Brunswick, Georgia. viii, 166 pp. A description of the fishes of Texas with range and keys. Illustrated.

Koelz, Walter. 1929. Coregonid Fishes of the Great Lakes. Bull. U.S. Bur. Fish., Vol. 43, 1927; Pt. 2: pp. 297-643. Detailed descriptions of all the coregonid fishes inhabiting the Great Lakes with illustrations.

Koelz, Walter. 1931. Coregonid Fishes of Northeastern North America. Papers Mich. Acad. Sci., Arts and Letters, Vol. 13, 1930: pp. 303-432. Detailed descriptions of the coregonid fishes found in waters other than the Great Lakes in northeastern North America. Very useful despite the fact that many species described are no longer valid or are recognized as varieties.

Koster, William J. 1957. Guide to the Fishes of New Mexico. Univ. of New Mexico Press and Dept. of Game and Fish. 116 pp.

Kuhne, Eugene R. 1939. A Guide to the Fishes of Tennessee and the Mid-south. Tenn. Dept. Cons., Nashville. 124 pp. Keys and descriptions for many of the fishes found in Tennessee. Illustrated.

Lachner, E.A., C.R. Robins, and W.R. Courtenay, Jr. 1970. Exotic Fishes and Other Aquatic Organisms Introduced into North America. Smithsonian Contributions to Zool. No. 59; 29 pp. Detailed information on introductions of fishes, giving time and sources of introduction.

LaMonte, Fancesca, 1945. North American Game Fishes. Doubleday Doran, Graden City. xiv, 202 pp. Descriptions and ranges of salt and freshwater game fishes with keys and record weights and lengths. Illustrated.

Legendre, Vianney. 1954. Volume 1. Key to Game and Commercial Fishes of the Province of Quebec. Game and Fish. Dept. Province of Quebec, Montreal. pp. 180. Illustrated and includes 101 species.

Massman, W.H. 1954. Marine Fishes in Fresh and Brackish Waters of Virginia Rivers. Ecology, Vol. 35; pp. 75-78.

McPhail, J.D. and C.C. Lindsey. 1970. Freshwater Fishes of Northwestern Canada and Alaska.

Fish. Res. Bd. Canada, Bull. 173: 381 pp. Detailed discussion of species found in northwestern North America, illustrations, distribution maps and keys.

Miller, R.J. and H.W. Robison. 1973. The Fishes of Oklahoma. Oklahoma State University Press, Stillwater, Oklahoma. 246 pp. Habits, distributions, descriptions and illustrations of Oklahoma fishes.

Miller, R.R. 1948. The Cyprinodont Fishes of the Death Valley System of Eastern California and Southwestern Nevada. Misc. Pub. Mus. Zool., Univ. Mich., No. 68, 155 pp.

Minckley, W.L. 1973. Fishes of Arizona. Arizona Game and Fish Dept., Phoenix; 293 pp. Descriptions of and keys to the native and introduced fishes of Arizona, with detailed distribution maps.

Moore, George A. 1968. Fishes. Pt. II; 22-165, in W.F. Blair et al, Vertebrates of the United States. 2nd ed. McGraw Hill Book Co., New York.

Morita, Clyde M. 1963. Freshwater Fishing in Hawaii. Div. of Game and Fish. Honolulu. 20 pp. A popular account of the introduced game and pan fishes in Hawaii.

Moyle, P.B. 1976. Inland Fishes of California. Univ. California Press, Berkeley, 405 pp. Descriptions of and keys to the native and introduced freshwater fishes of California, with detailed discussions of the biology, ecology and distribution of each species. Illustrated.

Paetz, M.J. and J.S. Nelson. 1970. The Fishes of Alberta. The Queen's Printer, Edmonton, 281 pp. Descriptions of and keys to the fishes of Alberta, including illustrations and range maps.

Pflieger, W.L. 1971. A Distributional Study of Missouri Fishes. Mus. Nat. Hist., Univ. Kansas Vol. 20: p. 225-570. Detailed distributional information with maps.

Pflieger, W.L. 1975. The Fishes of Missouri. Missouri Dept. of Cons. 343 pp. A semi-popular account of the fishes of Missouri, excellent keys, detailed distributional and biological information on the fish fauna of Missouri. Illustrated.

Rosen, D.E. and R.M. Bailey. 1963. The Poeciliid Fishes (Cyprinodontiformes), their Structure, Zoogeography and Systematics. Am. Mus. Nat. Hist. Bull. 126 (1); 176 pp. A detailed study of this group including those found in the United States.

Schultz, Leonard P. 1936. Keys to the Fishes of Washington, Oregon and Closely Adjoining Regions. Univ. Wash. Publ. Zool. Vol. 2, No. 4; pp. 103-228. Useful for identification of both marine and freshwater fishes of this region.

Scott, W.B. and E.J. Crossman. 1969. Checklist of Canadian Freshwater Fishes with Keys for Identification. Life Sci. Misc. Publ., Royal Ontario Mus. 104 pp.

Scott, W.B. and E.J. Crossman. 1973. Freshwater Fishes of Canada. Fis. Res. Bd. Canada, Bull. 184, 966 pp. A most detailed account of the fishes of Canada, with keys, figures, distribution maps. Much information on the natural history of the individual species.

Sigler, Wm. F., and R.R. Miller. 1963. Fishes of Utah. Utah State Dept. Fish and Game. 203 pp. Description and habits of the fishes of Utah.

Simon, James R. 1946. Wyoming Fishes. Bull. Wyo. Game and Fish Dept., Cheyenne. No. 4, 129 pp. Descriptions of Wyoming fishes with keys. Illustrated.

Slastenenko, E.P. 1958. The Freshwater Fishes of Canada. Toronto, Ont., 385 pp.

Smith, Hugh M. 1907. The Fishes of North Carolina. North Carolina Geol. and Econ. Surv. Vol. 2: xi—453 pp. Although out of date, still the most authoritative account of the fishes of this region.

Smith-Vanz, W.F. 1968. Freshwater Fishes of Alabama. Auburn Exp. Station, Auburn Univ., 211 pp. Key to the fishes of Alabama, discussion of distribution, illustrated.

Trautman, Milton B. 1957. The Fishes of Ohio, with Illustrated Keys. Ohio State Univ. Press, Columbus, 683 pp. An excellent description of the distribution and habits of the fishes of Ohio.

Walters, V. 1955. Fishes of Western Arctic America and Eastern Siberia. Bull. Am. Mus. Nat. Hist., 106(5): 225-368.

Whitworth, W.R., P.L. Berrien, and W. T. Keller. 1968. Freshwater Fishes of Connecticut. State Geol. and Nat. Hist. Surv. of Connecticut. Bull. 101, 134 pp.

Wilimovsky, Norman J. 1954. List of the Fishes of Alaska. Stanford Ichthyological Bull., 4(5): 279-294.

Zim, H.S. and H.H. Shoemaker. 1956. Fishes. Simon and Schuster, N.Y. 160 pp. A popular illustrated guide to 278 representative salt and freshwater fishes.

Index
and Pictured Glossary

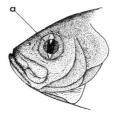
201

of the pectoral and/or pelvic
fins, it is surrounded by fat
and supported by a bony
splint in the Salmonidae. 7, 8,
19, figs. 31, 651(a), 656

Figure 651

B

BARBEL: thread-like structure
on the head, usually near
mouth, it may be minute and
at or near the end of the maxillary. Sensory in function in
many species. 11, 20, 23, 68,
134, figs. 37, 49, 201, 202, 653,
669(b)
BELLY: region on the underside
(ventral surface) of the fish
between the pectoral fin and
anus. 6, 7, figs. 1, 652(c)

BICUSPID: tooth with two
points. 26

BRANCHIAL CHAMBER: space
or cavity containing the gills.
11

BRANCHIOSTEGAL MEM-
BRANES: (gill membranes)
membranes extending below
the gill covers (opercles) and
connected with each other or
the isthmus (throat). 11,
fig. 652, 656
   RAYS: slender bones in the
   branchiostegal membranes.
   10, 11, fig. 9, 652(b)

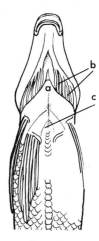

Figure 652

BREAST: (thorax) region under
and immediately before the
pectoral fins. 6, 7, fig. 1

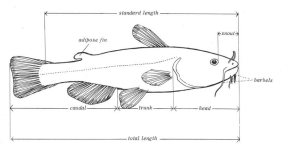

Figure 653

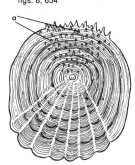
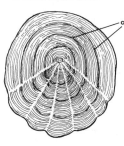

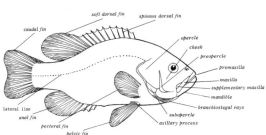

Figure 656

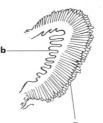

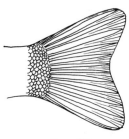

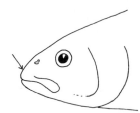

Figure 663

Figure 664

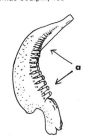
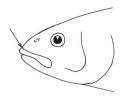

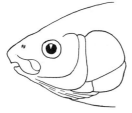

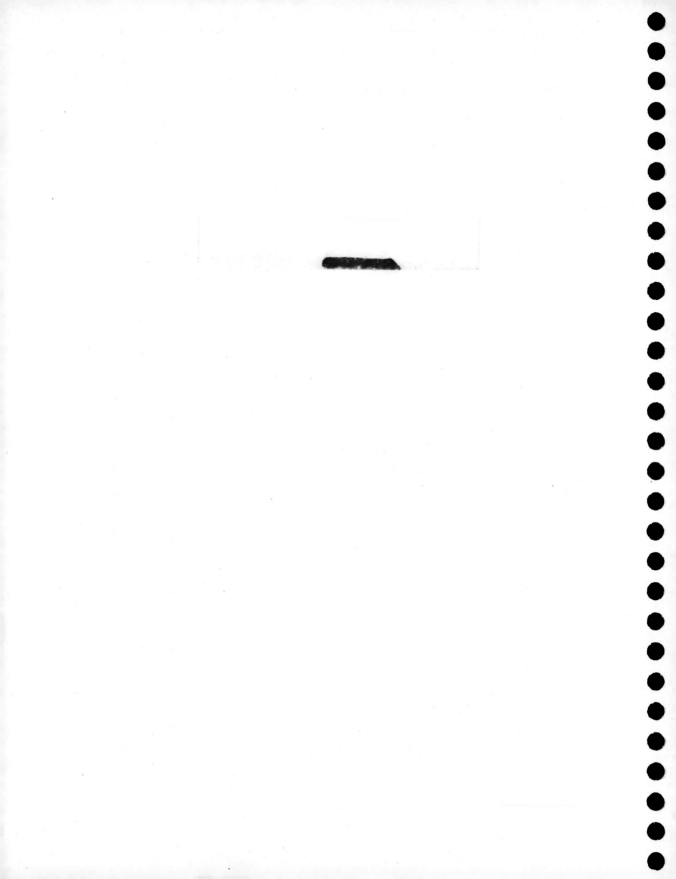